99-PUM-4

IMPACT OF SURFACE STORAGE ON RECLAIMED WATER: SEASONAL AND LONG TERM

by:

Gerald Miller, Ph.D.
Richard ten Bosch, P.E.
Alex Horne, Ph.D.
Andrew Lazenby
Elizabeth Quinlan
Carl Spangenberg, P.E.

2003

Water Environment Research Foundation

The Water Environment Research Foundation, a not-for-profit organization, funds and manages water quality research for its subscribers through a diverse public-private partnership between municipal utilities, corporations, academia, industry, and the federal government. WERF subscribers include municipal and regional water and wastewater utilities, industrial corporations, environmental engineering firms, and others that share a commitment to cost-effective water quality solutions. WERF is dedicated to advancing science and technology addressing water quality issues as they impact water resources, the atmosphere, the lands, and quality of life.

For more information, contact:
Water Environment Research Foundation
635 Slaters Lane
Alexandria, VA 22314
Tel: (703) 684-2470
Fax: (703) 299-0742
www.werf.org
werf@werf.org

This report was published by the following organization. For non-subscriber sales information, contact:

IWA Publishing
Alliance House, 12 Caxton Street
London SW1H 0QS, United Kingdom
Tel: +44 (0) 20 7654 5500
Fax: +44 (0) 20 7654 5555
www.iwapublishing.com
publications@iwap.co.uk

Library of Congress Catalog Card Number: 2002108105

WERF ISBN: 978-1-893664-52-4 IWAP ISBN: 978-1-84339-638-3

Black & Veatch

ACKNOWLEDGMENTS

Report Preparation

Principal Investigators:
Gerald Miller, Ph.D., *Black & Veatch*
Richard ten Bosch, P.E., *Black & Veatch*

Project Team:
Alex Horne, Ph.D., *University of California, Berkeley*
Andrew Lazenby, *Black & Veatch*
Elizabeth Quinlan, Project Scientist, *Black & Veatch*
Carl Spangenberg, P.E., *Irvine Ranch Water District*
Melissa Blanton

Technical Advisory Committee

Andrew Bath, Ph.D., *Ninham Shand Ltd., Republic of South Africa*
Alvin Smith, *Binnie Black & Veatch, United Kingdom*

Project Subcommittee

George Tchobanoglous, Ph.D., Chair, *Tchobanoglous Consulting*
Richard Sakaji, Ph.D., P.E., *California State Health Department*
Dann Yobbi, *U.S. Geological Survey*
Alan B. Cooper, P.E.
Jörg Drewes, Ph.D., *Colorado School of Mines*

Water Environment Research Foundation Staff

Research Program Director: Jami Montgomery
Project Manager: Bonnie Bailey

ABSTRACT

The objectives of this project were to develop 1) a better understanding of the effects of storage on reclaimed water quality, 2) a methodology to help understand/predict water quality changes during storage, and 3) effective management tools for minimizing water quality problems. The research team reviewed approximately 120 published articles, conducted a gray literature survey to analyze the impact of surface storage on reclaimed water quality. The team also evaluated federal guidelines for reclaimed water and developed a brief update on what individual states are doing.

It was determined that state and federal water quality objectives can be met at the treatment site. However, because of the seasonal nature of reclaimed water use, water often must be stored in open reservoirs, where changes occur that can affect water quality. The nature of these changes was evaluated, including physical, chemical, and biological processes.

The research team evaluated several reservoir management strategies to improve water quality, and reviewed water quality models to assess their applicability for open reclaimed water storage reservoirs. It also developed procedures to evaluate and select management strategies and reservoir water, along with matrices to distill the information learned in the study into a useful format for risk assessors and water quality managers. These tools will enable users to readily equate their specific storage reservoirs to representative examples, and to identify actions most applicable to their specific reclaimed water systems.

BENEFITS

This report provides valuable information of lake/reservoir water quality models and of management strategies for minimizing water-quality problems. It endeavors to enhance our understanding of what water quality computer modeling tools are available to effectively address water quality issues of seasonal and long-term surface storage of nonpotable reclaimed waters. It helps users equate their specific storage reservoir to representative examples in the document and identify a water quality computer model most applicable to their specific storage reservoir. It provides guidance to risk assessors and risk managers in identifying useful water quality parameters in storage facilities, the impact of both seasonal and long-term storage on water quality, and potential methods to both mitigate water quality deterioration and enhance water quality improvement.

- Provides useful summary of reclaimed water applications by state.
- Identifies water quality parameters affected by storage and the physical, chemical, and biological processes most important in effecting these changes.
- Identifies water management strategies that can be implemented to enhance water quality or to minimize water quality problems.
- Describes computer models available to simulate the water quality of open reclaimed water reservoirs.
- Provides information to enable risk assessors/managers to better identify useful water quality parameters in storage facilities, the impacts of seasonal and long-term storage on water quality, and potential methods to mitigate water quality deterioration and enhance water quality improvement.
- Improves the understanding of factors affecting water quality in reclaimed water reservoirs.
- Provides information on management techniques that can facilitate the control of water quality in reclaimed water reservoirs.

Keywords: Reclaimed water, nonpotable reuse, storage, management strategies, water quality models

TABLE OF CONTENTS

LIST OF TABLES

LIST OF FIGURES

LIST OF ACRONYMS

ADEQ	Arizona Department of Environmental Quality
AHOD	Areal Hypolimnetic Oxygen Demand
AWWA	American Water Works Association
BMPS	Best Management Practices
BOD	Biochemical Oxygen Demand
BOD_5	5-Day Biochemical Oxygen Demand
C	Carbon
CAEDYM	Computational Aquatic Ecosystem Dynamics Model
CBOD	Carbonaceous Biochemical Oxygen Demand
CCR	California Code of Regulations
CFU	Colony Forming Units
$C1_2$	Chlorine
CH_4	Methane
cm	Centimeter
cm/yr	Centimeters per Year
CO_2	Carbon Dioxide
CO_3^{-2}	Carbonate Ion
CUs	Color Units
CWA	Clean Water Act
CWR	Centre for Water Research
DHI	Danish Hydraulic Institute
DO	Dissolved Oxygen
DOE	Department of Ecology (State Of Washington)
DOH	Department of Health (State Of Washington)
DOHS	Department of Health Services (State Of California)
DYRESM	Dynamic Reservoir Simulation Model
ELCOM	Estuary, Lake, and Coastal Ocean Model
FDEP	Florida Department of Environmental Protection
H	Hydrogen
H_20	Water
H_2CO_3	Carbonic Acid
H_2S	Hydrogen Sulfide
ha	Hectares
HCO_3	Bicarbonate Ion
ILP	Irvine Lake Pipeline
IPR	Indirect Potable Reuse
IRWD	Irvine Ranch Water District
L	Lenath
M	Mass
m^3	Cubic Meters
MF	Microfiltration
mgd	Million Gallons per Day
mg/L	Milligrams per Liter

ml	Milliliters
MPN	Most Probable Number
MWRP	Michelson Water Reclamation Plant
N	Nitrogen
N_2	Molecular Nitrogen
NALMS	North American Lakes Management Society
NH_3	Ammonia
NH_4^+	Ammonium
NO_2^-	Nitrite
NO_3^-	Nitrate
NOD	Nitrogenous Oxygen Demand
NPDES	National Pollution Discharge Elimination System
NTU	Nephelometric Turbidity Unit
O_2	Oxygen
P	Phosphorus
PC	Personal Computer
PO_4^{3-}	Phosphate
PO_4^{3-} P/L	Phosphate as Phosphorus per Liter
SDWA	Safe Drinking Water Act
SMS	Surfacewater Modeling System
SO_4^{2-}	Sulfate
T	Time
TDS	Total Dissolved Solids
TIN	Total Inorganic Nitrogen
TKN	Total Kjeldahl Nitrogen
TOC	Total Organic Carbon
TOx	Total Organic Halogen
TNRCC	Texas Natural Resource Conservation Commission
TSS	Total Suspended Solids
TVA	Tennessee Valley Authority
U.S. EPA	United States Environmental Protection Agency
μg	Micrograms
μg/L	Micrograms Per Liter
UV	Ultraviolet Light
VOC	Volatile Organic Compounds
WEF	Water Environment Federation
WERF	Water Environment Research Foundation

Executive Summary

Research Objectives

In October 1999, WERF selected a research team led by Black & Veatch to perform Project 99-PUM-4. The objectives of the project are (1) to compile information on the effects of storage on reclaimed water quality, (2) to evaluate methods to help understand and predict water quality changes during storage, and (3) to identify effective management tools for minimizing water quality problems.

Literature Review

Literature review focused on two major tasks: literature search and gray literature survey. The literature search was used to identify a wide range of documents on water quality criteria, management strategies, and/or water quality models with application to open reclaimed water reservoirs. Approximately 120 documents were reviewed. An annotated bibliography of these documents is presented in Appendix A. The literature review also included development and implementation of a gray literature survey on the impact of surface storage on reclaimed water quality. The purpose of the survey was to obtain information on existing reclaimed water facilities and how they are managed. A copy of the survey and a summary of results are provided in Appendix B.

Water Quality Criteria

The project team reviewed categories of reclaimed water use, federal guidelines for various reuse applications, and current regulations of six selected states: California, Florida, Arizona, Nevada, Texas, and Washington. Currently, no specific federal regulations govern reclaimed water use; thus, quality requirements for reclaimed water are largely determined by customer requirements and by state regulations.

Reclaimed water can be reused for a number of applications. Treatment requirements range from conventional to advanced, depending upon how the water will be reused. Applications include agricultural irrigation (surface and spray), landscape irrigation, recreational use (restricted and nonrestricted), industrial reuse, urban uses, and groundwater recharge.

Federal guidelines for water reuse are contained in the US Environmental Protection Agency (U.S. EPA) 1992 guidance manual on reuse (U.S. EPA, 1992). Constituents of concern include suspended solids, biodegradable organics, nutrients, stable organics, hydrogen ion concentration, heavy metals, dissolved inorganics, and residual chlorine. Also included in the manual are suggested treatment and reclaimed water quality guidelines categorized by type of reuse.

Water quality standards for various types of irrigation uses for the six above states were analyzed and compared. Water quality standards for pasture irrigation, irrigation of restricted access areas, urban irrigation, and spray irrigation of food crops eaten raw are summarized. In general, most of the states require at least secondary treatment and disinfection for each irrigation use. Each state has specific water quality limits defined by total and fecal coliform, biochemical oxygen demand (BOD), and total suspended solids (TSS). For all six states, spray irrigation of food crops eaten raw has the most stringent water quality limits while pasture irrigation and irrigation of restricted access areas have the most relaxed limits.

Water Quality Parameters

Treated wastewater can be reused for a wide variety of applications, and treatment technologies are available to provide the needed water quality protections. However, water reuse projects are often seasonal in nature and require storage of reclaimed water for long periods. Frequently, reclaimed water is held in open storage reservoirs, and, during periods of storage, the quality of the water can change.

Many of the significant water quality parameters that impact the beneficial uses of reclaimed water reservoirs are the same as for freshwater reservoirs: physical, chemical, biological, and aesthetic parameters. Also, the quality of reclaimed water in reservoirs can change through constituent transformation and removal processes, or fate processes. Perhaps the greatest difference is that the water supply for reclaimed water reservoirs can have significantly higher levels of nutrients (phosphorous and nitrogen). Differences in nutrient levels have a direct effect on biological uptake and alga responses in reclaimed water reservoirs.

Reclaimed water reservoirs, typically smaller than freshwater reservoirs, are used primarily for operational purposes, as opposed to flood control storage that occurs during storm events. During the off peak demand periods of fall and winter, reclaimed water reservoirs are filled, while maintaining a certain freeboard for runoff. However, during the summer, when reclaimed water use is the greatest, they are used to manage daily fluctuations in demand.

Reclaimed water reservoirs follow the same traditional thermal patterns within the water column of stratification/destratification as freshwater reservoirs, resulting in nutrient cycling and algae growth responses. However, due to excessive levels of nutrient availability, traditional biological responses and relationships found in freshwater systems cannot be used.

Irvine Ranch Water District (IRWD) in Southern California operates two open-storage reclaimed water reservoirs, Rattlesnake and Sand Canyon Reservoirs, which were selected for a case study. Water quality problems experienced at the reservoirs include turbidity, odor, and algae growth. Water quality degradation was more severe in Sand Canyon because the reservoir is more shallow than Rattlesnake, has a larger watershed, and collects fine clays during storm events. As a result of these problems, IRWD implemented a watershed management plan.

Reservoir Management Strategies

Reservoir management includes a set of strategies designed to improve water quality, and a number of these are currently used in lakes and freshwater reservoirs. These strategies may be applicable to open reclaimed water reservoirs, but it is important to note that reclaimed water reservoirs are significantly different from natural lakes, and such differences must be carefully considered when reviewing results of programs conducted elsewhere. The effectiveness of any management strategy will depend to a large extent on site-specific conditions, including the physical characteristics of the reservoir, runoff from the watershed, the initial quality of the reclaimed water, and reservoir operations.

To evaluate and select strategies that could be effective in improving water quality in their reservoirs, utility managers can take the following approach: identify problems; define the physical and operational characteristics of the reservoir; screen potential strategies; evaluate alternatives; and select feasible management strategies.

For Project 99-PUM-4, four strategies were dropped from consideration after a preliminary screening: hypolimnetic withdrawal, dilution, biomanipulation, and targeted plant growth. These were categorized as unlikely to be effective in improving water quality in reclaimed water storage reservoirs. Nine strategies given more detailed evaluation were:

- destratification/artificial circulation
- hypolimnetic aeration/oxygenation
- wetlands treatment
- copper treatment
- advanced waste treatment for phosphorus removal
- advanced waste treatment for nitrogen removal
- dredging
- alum treatment
- watershed management

Evaluation of the strategies is summarized in a matrix table listing problems addressed, effectiveness, and other considerations.

Reservoir Water Quality Models

Most computer models applicable to reclaimed water reservoirs were developed for natural lakes or reservoirs. Many focus on eutrophication, which is also an issue for reclaimed water reservoirs. Types of available models include empirical, process, steady-state, and dynamic. Because of the range of facilities, operations, and end uses of reclaimed water, no one model emerges as the “best” for reclaimed water reservoirs. Water utility managers can determine the model most appropriate for their needs by considering three factors: (1) the ability of the model to simulate conditions present in reclaimed water reservoirs, (2) its effectiveness in addressing the issues of key water quality parameters, water quality criteria, and management strategies, and (3) the requirements and limitations of the model. Using these three factors, the Project Team evaluated 12 models, listed in Table ES-1.

The evaluation of the water quality models is summarized in a matrix table (Table 6-1) listing model type, parameters simulated, processes simulated, complexity, data and user requirements, and other factors.

Management Tools

Typical water quality problems encountered in reclaimed water storage reservoirs are excess nutrients and algae, low dissolved oxygen (DO), high turbidity, odors, and color. These problems do not occur independently, but are interrelated. Management strategies are designed to either address the cause of the problems, i.e., the various sources of nutrients, or treat the water quality problems directly.

With a water quality model, managers can estimate the extent to which each source of nutrients should be controlled. Most of the models reviewed are capable of examining some aspects of nutrient reduction strategies. The ability of the various models to simulate water quality problems and management strategies is summarized in Table ES-1.

Table ES-1. Water Quality Models Applicable to Water Quality Problems and Management Strategies

Model	Type	Water Quality Problems				Management Strategies				
		Nutrients	Low DO	Algae	Other	Reduce Nutrient Loads			Low DO	Algae
						Reclaimed Water	Watershed Runoff	Internal Sources		
EUTROMOD	well-mixed, steady state, empirical	Y		B		Y	Y			
PHOSMOD	well-mixed, steady state, empirical	Y	B			Y	Y	B		
BATHTUB	well-mixed segments, steady state, empirical	Y	B	B		Y	Y			
CE-QUAL-R1	1-D (vertical), dynamic, process	Y	M	M	Y	Y	Y		B	B
MIKE 21	2-D (horizontal), dynamic, process	Y		M	Y	Y	Y			M
RMA4	2-D (horizontal), dynamic, process				Y					
MIKE 11 RESERVOIR	2-D (vertical, longitudinal), dynamic, process	Y	A	M	Y	Y	Y	M	A	M
CE-QUAL-W2	2-D (vertical, longitudinal), dynamic, process	Y	A	M	Y	Y	Y	M	A	M
WASP5	3-D, external dynamic model, process	Y	A	M	Y	Y	Y	A	A	M
CE-QUAL-ICM	3-D, external dynamic model, process	Y	A	A	Y	Y	Y	A	A	A
MIKE 3	3-D, dynamic model, process	Y	A	M	Y	Y	Y	A	A	M
CAEDYM	3-D, external dynamic model, process	Y	A	A	Y	Y	Y	A	A	A

A = Advanced ability to address water quality problems and management strategies.
B = Basic ability to address water quality problems and management strategies.
M = Medium ability to address water quality problems and management strategies.
Y = Capable of simulating nutrient cycles and evaluating external nutrient reductions.

The study included (1) a review of existing information on management strategies that address water quality issues in open reclaimed water reservoirs and (2) a review of existing models that could simulate water quality problems and the effectiveness of applying various management strategies to a reservoir with water quality problems. These reviews are completed in Chapters 5.0 and 6.0. The information is synthesized in Chapter 7.0 so that utility managers can follow a logical path to determine which management strategy may successfully address their water quality problem(s) and which water quality model may be useful to understand specific problems and simulate the effects of management strategies.

CHAPTER 1.0

INTRODUCTION

1.1 Overview

In October 1999, the Water Environment Research Foundation (WERF) selected a research team led by Black & Veatch to perform Project 99-PUM-4, *Impact of Surface Storage on Reclaimed Water for Nonpotable Use: Seasonal and Long Term*. The results of that study are presented in this report. Project objectives and methodology are discussed below.

1.2 Objectives

Storage of reclaimed water is used to balance production and demand in water supply systems and can be long-term or seasonal, depending on the method of system operation. Because storage in open reservoirs changes the characteristics of reclaimed water and has potential water quality impacts, a better understanding of the effects of storage on reclaimed water quality is needed. Also required is a methodology to help understand/predict water quality changes during storage and to identify effective management tools for minimizing water quality problems. The overall objective of this project was to provide a useful tool to understand and manage reclaimed water facilities. Specific objectives were to:

- Compile information on computer models available to simulate the water quality of reclaimed water storage reservoirs. Identify the applicability of these models and model limitations relative to reclaimed water.
- Identify the water quality parameters affected by storage and the physical, chemical, and biological processes most important in effecting these changes. Identify reuse water quality criteria.
- Identify management strategies that can be implemented to enhance water quality or to minimize water quality problems.

The material presented should enable water utility managers, treatment plant operators, and other resources personnel to better identify useful water quality parameters in open storage facilities, the impact of both seasonal and long-term storage on water quality, and potential methods to mitigate water quality deterioration and enhance water quality improvement. In addition, they should also be able to readily equate their specific storage reservoirs to representative examples described herein and to identify the water quality computer models most applicable to their specific reclaimed water systems.

1.3 Participants

In addition to Black & Veatch investigative staff and technical specialists, a number of individuals and agencies assisted on the research and/or peer review. As described more fully in subsequent chapters, Water Quality Division staff from Irvine Ranch Water District (IRWD) helped on a number of tasks. IRWD is a pioneer in water recycling and has a wealth of knowledge, including operational data from two open storage reclaimed water reservoirs: Rattlesnake and Sand Canyon. Mr. Carl Spangenburg, Senior Engineer, led the IRWD effort. Dr. Alex Horne, Professor of Ecological Engineering, Department of Civil and Environmental Engineering, University of California, Berkeley, also participated on the team.

Seventeen public agencies participated in a gray literature survey on the impact of surface storage on reclaimed water, seasonal and long term. These agencies are Buena Sanitation District, Dublin San Ramon Services District, Eastern Municipal Water District, Irvine Ranch Water District, Las Virgenes Municipal Water District, County of Los Angeles Sanitation Districts, Monterey Regional Water Pollution Control Agency, Orange County Pubic Utilities, City of Phoenix Water Service Department, Pinellas County Utilities, Santa Clara Valley Water District, Santa Margarita Water District, South Bay Water Recycling Program, Tucson Water Department, Vallecitos Water District, City of Vernon, and City of Westminster.

In addition, a WERF-selected technical panel also reviewed the work at key project milestones. The WERF Project Subcommittee is chaired by Dr. George Tchobanoglous, Professor Emeritus, University of California, Davis. The other subcommittee members are Mr. Alan B. Cooper, Metcalf & Eddy; Dr. Jörg Drewes, Colorado School of Mines; Dr. Richard Sakaji, California Department of Health Services; and Mr. Dan Yobbi, U.S. Geological Survey. Ms. Bonnie Bailey is the WERF Project Manager. The technical panel participated in a February 2001 workshop and reviewed progress reports and the December 2001 Draft Report. Comments of the panel are incorporated in this Final Report.

1.4 Methodology

Black & Veatch developed a research approach based on six specific tasks: Literature Review, Water Quality Criteria, Water Quality Parameters, Reservoir Management Strategies, Reservoir Water Quality Models, and Water Quality Management Tools. These tasks are summarized in Table 1-1 and in the narrative paragraphs that follow. More detailed information is provided in Chapters 2.0 through 7.0.

Table 1-1. Research Approach

Task	Objective	Approach
Literature Review • Literature Search • Survey	Compile information on water quality models, criteria, parameters, and reservoir management strategies Identify types of storage used, water quality problems, and current modeling effects.	• Utilize search tools including COMPENDEX, Applied Science and Technology, Environment, and WORLDCAT. • Conduct additional searches using professional organizations and government agency web sites. • Develop and distribute survey to reclaimed water utilities. • Analyze data and conduct follow-up interviews.
Water Quality Criteria	Identify water quality criteria pertinent to reuse.	• Review federal guidelines and regulations of six states to identify the important criteria.
Water Quality Parameters	Identify water quality parameters affected by storage.	• Evaluate physical, chemical, and biological processes; fate process and transformations; and aesthetic parameters.
Reservoir Management Strategies	Identify management strategies that could mitigate water quality problems found in reclaimed water reservoirs.	• Review management strategies and eliminate strategies with limited potential for storage of reclaimed water. • Evaluate remaining strategies for problems addressed and effectiveness for reclaimed water reservoirs.
Reservoir Water Quality Models	Identify models that could be a useful tool in managing reclaimed water reservoirs.	• Review existing water quality models for effectiveness with respect to storage of reclaimed water. • Evaluate the parameters and processes simulated, degree of complexity, and data and user requirements.
Water Quality Management Tools	Provide a useful tool to allow utilities to better understand and manage reclaimed water facilities.	• Provide information in a useful format to assist reclaimed water managers select models and management strategies.

1.4.1 Literature Review

The literature review was divided into two major subtasks: literature search and gray literature survey. To conduct the literature review, Black & Veatch utilized several search tools, including COMPENDEX, Applied Science and Technology, Environment, and WORLDCAT. In addition, the research team searched several websites. An annotated bibliography of approximately 120 documents was prepared and is included in Appendix A. The second part of the literature review included development, conduct, and analysis of the results of a survey on the impact of surface storage on reclaimed water, seasonal and long term. Sixteen public agencies participated in the survey. Survey information is presented in Appendix B.

1.4.2 Water Quality Criteria

The project team evaluated federal guidelines for reclaimed water used to augment surface water supplies, including Safe Drinking Water Act (SDWA) and Clean Water Act (CWA) requirements. Guidelines and criteria issued by selected states were also reviewed. Issues, principles, and recommendations found in the literature when the normal controls of the CWA and SDWA become insufficient were also reviewed.

1.4.3 Water Quality Parameters

This task involved identification of the processes that occur to change water quality in open storage reservoirs and the key water quality parameters that are affected. Physical, chemical, and biological processes were included. Fate processes and transformations and aesthetic parameters were also reviewed. A case study was developed for two Southern California reclaimed water storage reservoirs.

1.4.4 Reservoir Management Strategies

Reservoir management includes a set of strategies designed to improve the quality of water in the reservoir. Several strategies were evaluated as a part of this task. During a preliminary screening, the project team eliminated hypolimnetic withdrawal, dilution, biomanipulation, and targeted plant growth. A more detailed evaluation was made of destratification/artificial circulation, hypolimnetic aeration/oxygenation, wetlands treatment, copper treatment, advanced waste treatment for both phosphorus and nitrogen removal, dredging, alum treatment, and watershed management.

1.4.5 Reservoir Water Quality Models

Most computer models that would be applicable to reclaimed water storage reservoirs were developed for natural lakes or reservoirs. While it may be possible to use these models for reclaimed water, it is important to verify the model results for conditions that are beyond the values used in the development of the models. The 12 models reviewed as part of the project included empirical, process, steady-state, and dynamic types. Evaluated models were EUTROMOD, PHOSMOD, BATHTUB, CE-QUAL-R1, CE-QUAL-W2, RMA2 and RMA4, MIKE 11 Reservoir, Mike 21, CE-QUAL-ICM, WASP5, MIKE 3, and Computational Aquatic Ecosystem Dynamics Model (CAEDYM).

1.4.6 Water Quality Management Tools

The information developed in the previous tasks was presented in a format designed to assist reclaimed water utilities in accomplishing the following objectives:

1. To identify key characteristics of their reservoirs and operation that are important in making comparisons to other facilities and in model selection.
2. To identify a model that could be a useful tool in managing their facilities.
3. To understand the cause of their water quality problems and identify management strategies that may mitigate the problem.

CHAPTER 2.0

LITERATURE REVIEW

2.1 Overview

Literature review focused on two major subtasks: literature search and gray literature survey. The literature search was conducted to identify available information on water quality parameters, water quality criteria, management strategies, and water quality models that may be appropriate for modeling reclaimed water reservoirs. The second subtask involved developing, conducting, and evaluating results of a gray literature survey to compile unpublished information on research performed by numerous municipalities, reclaimed water utilities, and other organizations. The results of the literature review and the gray literature survey are presented in Appendix A and Appendix B, respectively. Key facts are summarized below.

2.2 Literature Search

To conduct the literature review, Black & Veatch utilized several search tools including COMPENDEX, Applied Science and Technology, Environment, and WORLDCAT. In addition, the research team searched several web sites, including the Environmental Protection Agency (www.epa.gov), the American Water Works Association (www.awwa.org), the Water Environment Federation (www.wef.org), and the United States Geological Survey (www.usgs.gov). The annual literature review provided by the Water Environment Federation (WEF) also provided references to many additional reports. These documents were also reviewed and information/data compiled.

Approximately 120 documents are included in the bibliography for this project. An annotated bibliography was prepared and is included in Appendix A.

The general references included information in more than one category. These documents typically included discussions on limnological processes and important water quality parameters, lake restoration or management strategies, and basic information on modeling. Many other references were reviewed related to water quality models and management strategies. Most of these are not specifically applicable to reclaimed water reservoirs, and therefore, information has been extrapolated for use with reclaimed water reservoirs. The information obtained from the documents is discussed in Chapters 3.0 through 6.0.

2.3 Gray Literature Survey

The second part of the data review involved conducting a gray literature survey on the *Impact of Surface Storage on Reclaimed Water, Seasonal and Long Term*. A copy of the survey and a discussion of results are provided in Appendix B and summarized below.

2.3.1 Survey Development

Irvine Ranch Water District (IRWD) Water Quality Division staff, with assistance from Black & Veatch, developed the survey. Its purpose was to obtain information on reclaimed water facilities and how they are managed. Questions focused on:

- The type of facilities used by the utility.
- Storage requirements and primary purpose of storage.
- End uses of the reclaimed water.
- Water quality problems encountered and management strategies, if any.
- Monitoring program and type of data collected.

The survey was sent to municipalities, reclaimed water utilities, and other organizations that may use surface storage of reclaimed water for non-potable uses. The mailing list was derived from the 1999/2000 American Water Works Association (AWWA) Water Reuse Database, which lists agencies that utilize or are planning to utilize reclaimed water. Based on information gained by project team members through participation on the AWWA Water Reuse Committee, 20 agencies were contacted by phone to verify their willingness to participate in the survey. A package of information was prepared and faxed or e-mailed to each participant. It included a brief description of the project, a list of key project investigators, and the survey form. Follow-up phone calls were made to each participant to verify the information was received and to walk through each survey question to ensure the responder understood the information being requested.

2.3.2 Survey Responses

Seventeen agencies responded to the survey, 11 of which had surface storage of reclaimed water for non-potable use. These initial responses indicated that aeration was the most commonly used water quality management strategy and that hypolimnetic withdrawal was the second.

All survey respondents with open reclaimed water reservoirs conducted water quality monitoring at the treatment plant effluent, and some agencies also monitored water quality at their distribution systems and/or storage reservoirs. The parameters monitored were typically turbidity, suspended solids, chlorine residual, and total coliform count.

The most frequent approaches to water quality sampling were to take samples weekly and/or to take samples in response to water quality problems. Algae blooms were the most common water quality problem experienced.

2.3.3 Follow-Up Contact

Upon compiling the results from the gray literature survey, the research team determined that five of the agencies operating open reclaimed water reservoirs should be contacted to obtain additional information. Each was asked the following questions:

- During our initial survey, your agency indicated that *[management strategy]* is used at *[name of reservoir]* to control *[water quality problem(s)]*. Has this strategy been successful?
- Has any other management strategy been used in the past to address this problem? If so, what was that strategy, and why is it no longer used?
- Has your agency ever used a computer model to determine potential water quality management strategies experienced at your facility? If so, what is the name of the model?

In general, most agencies indicated that the management strategy currently being used has been effective in maintaining acceptable water quality; however, few of the strategies have actually eliminated the problems. Two agencies indicated that copper sulfate was used in the past to eliminate algae blooms. Both indicated that, although the copper sulfate treatment solved the problem, the process was no longer considered a feasible solution because it is very labor intensive. None of the five reported having ever used a computer model to determine the most effective management strategy for their storage facilities. Four of the agencies were unaware of the availability of such models; the fifth noted that, in general, the models currently available are too expensive for its particular application.

CHAPTER 3.0

WATER QUALITY CRITERIA

3.1 Overview

The focus of the following discussion is on federal and state guidelines that address water quality criteria for reclaimed water used for a variety of nonpotable applications. Currently, no federal regulations specifically govern water reuse. Thus, quality requirements for reclaimed water are largely determined by customer requirements, for example, the type of water quality the customer desires, and possibly by state regulations. Although no formal federal requirements exist for reclaimed water, the United States Environmental Protection Agency (U.S. EPA) has developed guidelines. In addition, most states have developed regulations for reclaimed water use. Federal guidelines, state regulations, and customer requirements are largely driven by the intended use of the reclaimed water. The various categories of reclaimed water for nonpotable use, federal guidelines for water reuse, and water quality criteria recently developed by six selected states—California, Florida, Arizona, Nevada, Texas, and Washington—are presented and discussed in this chapter. These six states were selected as representative of the 36 states that have regulations or guidelines for reclaimed water.

3.2 Categories of Reclaimed Water Use

Reclaimed water can be reused for a number of applications. Treatment capabilities range from basic to advanced, depending upon how the water will be reused. Types of use, treatment guidelines, and reclaimed water quality standards are defined differently by different states. However, in general, reclaimed water uses can be divided into the seven major categories described below.

3.2.1 Agricultural Irrigation (Surface and Spray)

Traditionally, the greatest percentage of reclaimed water use has long been for agricultural applications. Uses include surface irrigation of orchards and vineyards and surface irrigation of fodder, fiber, and seed crops. Other agricultural applications, for which reuse requirements are generally more restrictive, are surface irrigation of food crops and surface irrigation of pastures for milking animals. The most restrictive requirements are generally reserved for spray irrigation of food crops.

3.2.2 Landscape Irrigation

Landscape irrigation provides another opportunity for reclaimed water use. In fact, some states, such as California, prohibit the use of potable water for landscape irrigation when reclaimed water of "suitable" quality is available. Landscape irrigation uses include golf courses, cemeteries, and roadway and freeway medians. Reclaimed water can also be used for landscape irrigation of parks and playgrounds although requirements are generally more stringent for these applications.

3.2.3 Recreational Use (Restricted and Nonrestricted)

Use of reclaimed water for recreational applications is less common although some projects have been in existence for a number of years. Restricted recreational impoundments, such as an ornamental lake on a golf course or landscaped "water features," generally must meet less stringent requirements than unrestricted impoundments, where no limitations are imposed on body-contact water recreational activities.

3.2.4 Industrial Reuse

Industries are a growing sector for reclaimed water reuse. A high level of treatment is generally required to meet industrial process needs. However, many successful examples point to the continued potential for this reuse, particularly for industries that require large amounts of process water, such as oil refineries and electronics manufacturers.

3.2.5 Urban Uses

Urban uses include ornamental fountains and other water features, commercial uses (car washes, air conditioning chiller water, etc.), street cleaning, and residential irrigation. Some pioneering work is also being done by local water districts in the use of dual water systems in newly constructed buildings to provide reclaimed water for flushing toilets and urinals.

3.2.6 Groundwater Recharge

Planned use of groundwater recharge with reclaimed water is not common, although some notable projects have shown the potential for this application. Different types of feed water, including reclaimed water, are used in groundwater recharge projects in California. Utilities in Arizona are using only reclaimed water for groundwater recharge applications. The type of groundwater recharge – surface spreading or direct injection—determines the treatment standard (Natural Resources Council, 1998).

3.2.7 Surface Water Augmentation

Although the focus of this research project is on non-potable reuse, it should be noted that another potential reclaimed water application is for augmentation of surface water supplies, where reclaimed water would be injected directly into a drinking water source. Surface water augmentation projects proposed to date would put reclaimed water in a raw water reservoir to mix with the water supply source prior to the mix being treated at a conventional water treatment plant. To date, planned surface water augmentation has been evaluated only on a test basis.

Public health issues and public perception issues must be addressed before this application becomes common practice.

It is clear that municipal wastewater is not the most desirable source water for potable purposes. Untreated municipal wastewater contains significantly higher concentrations of both microbial and chemical contaminants than natural protected sources of supply. The federal Clean Water Act (CWA) is intended to eliminate pollution and to maintain the chemical, physical, and biological integrity of the nation's waters; its water quality limits are not reflective of standards for reclaimed water. Similarly, the Safe Drinking Water Act (SDWA) requirements assure a safe drinking water when relatively uncontaminated, protected water sources are used. Drinking water regulations are not intended to address the issues raised by the reuse of reclaimed water.

Even when the contribution of wastewater discharges to a drinking water source exceed some threshold compliance with CWA and SDWA requirements, the provisions of these acts are insufficient to address all the public health concerns with municipal wastewater contaminants. The reason that some concerns are not addressed is because neither the CWA nor the SDWA establish standards for all harmful contaminants present in wastewater. The level of wastewater contribution that triggers additional contaminant controls has not been identified in law or in the indirect potable reuse (IPR) literature. The threshold level in any particular case depends upon a number of factors including: (1) the industrial, commercial, research, and medical contributions to the sewage that may present unique problems and (2) the natural barriers to contaminant transport that exist between the waste discharge(s) and the use of the reclaimed water.

With increased pressure to utilize dwindling water resources, greater emphasis is being placed on the conservation of existing supplies and on the use of reclaimed water with appropriate safeguards.

3.3 Federal Guidelines

U.S. EPA's guidance manual on reuse (U.S. EPA, 1992) is not a formal regulatory document, but it does provide recommendations for a wide range of reuse practices. Pertinent portions of the manual are summarized below.

3.3.1 Constituents of Concern

The U.S. EPA guidance manual provides a good discussion of inorganic and organic constituents of concern in water reclamation and reuse. These concerns are summarized in Table 3-1.

Other useful parameters include coliform bacteria and turbidity. Coliforms can be used as an indicator of the presence of pollution and possible pathogenic microorganisms in the water. Turbidity can be used as a measure of the clarity of the water. High turbidity can create aesthetic problems and can adversely affect disinfection of the reclaimed water.

3.3.2 Suggested Guidelines for Reuse

The guidelines presented in the manual provide general recommendations on the type of treatment and likely water quality requirements. They were not meant to be used as definitive water reclamation criteria, but only as reasonable guidance, particularly in states that had not developed their own criteria or guidelines. Suggested guidelines are summarized in Table 3-2, which is derived from a much longer table in the manual.

Table 3-1. Inorganic and Organic Constituents of Concern in Water Reclamation and Reuse*

Constituent	Measured Parameters	Reason(s) for Concern
Suspended Solids	Suspended Solids (SS) and Total Suspended Solids (TSS) including volatile and fixed solids.	♦ Possible plugging in irrigation systems. ♦ Potential to shield microorganisms from disinfectants.
Biodegradable Organics	Biochemical oxygen demand (BOD) Chemical oxygen demand Total organic carbon (TOC)	♦ Potential aesthetic and nuisance problems ♦ Potential adverse effects on disinfection process and/or industrial treatment processes
Nutrients	Nitrogen Phosphorus Potassium	♦ Possible promotion of growth of undesirable aquatic life ♦ Potential for nitrate build-up in groundwater
Stable Organics	Specific compounds such as pesticides, chlorinated hydrocarbons	♦ Potential resistance to conventional treatment ♦ Potentially toxic to the environment
Hydrogen Ion Concentration	pH	♦ Potential effect on disinfection process
Heavy Metals	Specific elements (such as cadmium and zinc)	♦ Potentially toxic to plants and animals
Dissolved Inorganics	Total Dissolved Solids (TDS) Specific elements	♦ Potentially toxic to crops
Residual Chlorine	Free and combined chlorine	♦ Potentially damaging to crops ♦ Potentially toxic to groundwater

*Adapted from U.S. EPA, 1992.

The recommended level of treatment depends on the type of reuse and can be site-specific. Both treatment processes and water quality limits were recommended by U.S. EPA (1992) because the combination of requirements known to produce reclaimed water of acceptable quality makes the monitoring of finished water for certain constituents unnecessary. In addition, U.S. EPA has indicated that it would be impractical to monitor reclaimed water for all of the chemical constituents and pathogenic organisms of concern and that surrogate parameters could be used because they are universally accepted. As mentioned previously, total and fecal coliforms are the most commonly used indicator organisms for reclaimed water. U.S. EPA considered fecal coliforms, however, to be a better indicator of fecal contamination and suggested that they be the surrogate to be used for monitoring.

Limits for parasites or viruses are not addressed in the guidelines. At the time, parasites had not been a problem at water reuse operations in the U.S. at the treatment levels and water quality limits recommended in the guidelines. It was recognized that viruses were a concern in reclaimed waters, but virus limits were not recommended for the following reasons:

- The difficulty in and high costs of identifying and counting viruses in wastewater.
- The limited number of facilities with the capability and staff to conduct the analyses.
- The extensive length of time to culture and identify viruses.
- The existence of a significant body of literature indicating that viruses were reduced or inactivated to low or immeasurable levels by appropriate wastewater treatment.

Table 3-2. U.S. EPA Suggested Guidelines for Water Reuse
(U.S. EPA, 1992)

Type of Use	Treatment	Reclaimed Water Quality
Urban reuse(landscape irrigation, fire protection, etc.), agricultural reuse (crops not processed), recreational impoundments (full body contact)	♦ Secondary ♦ Filtration ♦ Disinfection	♦ pH = 6 – 9≤10 mg/L BOD ♦ ≤2 NTU[a] ♦ No detectable fecal coli/100 mL[b] ♦ ≥1 mg/L Cl_2 residual[c]
Restricted access area irrigation, agricultural reuse (processed food crops and nonfood crops), aesthetic impoundments	♦ Secondary ♦ Disinfection	♦ pH = 6 – 9≤30 mg/L BOD ♦ ≤30 mg/L TSS ♦ ≤200 fecal coli/100 mL[e] ♦ ≥1 mg/L Cl_2 residual[c]
Construction uses, industrial cooling[d], environmental reuse	♦ Secondary ♦ Disinfection	♦ ≤30 mg/L BOD ♦ ≤30 mg/L TSS ♦ ≤200 fecal coli/100 mL[e] ♦ ≥1 mg/L Cl_2 residual[c]
Groundwater recharge of nonpotable aquifers by spreading	♦ Site-specific and use dependent ♦ Primary (minimum)	♦ Site-specific and use dependent
Groundwater recharge of nonpotable aquifers by injection	♦ Site-specific and use dependent ♦ Secondary (minimum)	♦ Site-specific and use dependent
Groundwater recharge of potable aquifers by spreading	♦ Site-specific ♦ Secondary & Disinfection (minimum)	♦ Site-specific ♦ Meet drinking water standards after percolation through vadose zone
Groundwater recharge of potable aquifers by injection, augmentation of surface supplies	Includes, but not limited to, the following: ♦ Secondary ♦ Filtration ♦ Disinfection ♦ Advanced wastewater treatment	Includes, but not limited to, the following: ♦ pH = 6.5 – 8.5 ♦ ≤2 NTU[a] ♦ No detectable fecal coli/100 mL[b] ♦ ≥1 mg/L Cl_2 residual[c] ♦ Meet drinking water standards

[a] Should be met prior to disinfection. Average based on a 24-hour time period. Turbidity should not exceed 5 NTU at any time.
[b] Based on seven-day median value. Should not exceed 14 fecal coli/100 mL in any sample.
[c] After a minimum contact time of 30 minutes.
[d] Recirculating cooling towers.
[e] Based on seven-day median value. Should not exceed 800 fecal coli/100 mL in any sample.

U.S. EPA related the removal of suspended material to disinfection capability. It was noted that many pathogens are particulate-associated and that particulate matter could shield both bacteria and viruses from disinfectants. U.S. EPA stated that it was generally accepted that particulate matter should be lowered to reduced levels [for example, 5 milligrams per liter (mg/L) TSS (total suspended solids) or 2 nephelometric turbidity units (NTU)] prior to disinfection. Continuously monitored turbidity was judged to be superior to composited TSS samples as an aid to wastewater treatment operations.

U.S. EPA related the removal of organic matter to the type of reuse. Some of the adverse impacts of organic substances are odor and color. Organic matter is also a food source for microorganisms, which can adversely affect the disinfection processes and lower dissolved oxygen levels (DO) when they decompose. U.S. EPA intended that the recommended BOD limit would indicate that the organic matter had been stabilized and was lowered to levels commensurate with anticipated types of reuse. BOD and TSS limits were suggested as an indicator of organic and particulate matter in reclaimed water that had received secondary treatment.

3.4 State Criteria

The U.S. EPA guidance manual presents an extensive review of state reuse regulations and guidelines. Appendix A of the manual contains eight tables outlining regulations/guidelines for unrestricted urban reuse, restricted urban reuse, agricultural reuse (food crops), agricultural reuse (non-food crops), unrestricted recreational reuse, restricted recreational reuse, environmental – wetlands, and industrial reuse. The information in the eight Appendix A tables was summarized by Crook (1998) and is presented in Table 3-3.

Table 3-3. Summary of State Water Reuse Regulations and Guidelines
(Adapted from Crook, 1998)

State	Regulations or Guidelines	No Regulations or Guidelines	Urban Reuse		Agricultural Reuse		Recreational Reuse		Environ-mental Reuse	Indus-trial Reuse
			Unrestricted	Restricted	Food Crops	Nonfood Crops	Unrestricted	Restricted		
AK		•								
AL						•				
AR	•		•	•	•	•				
AZ	•		•	•	•	•	•	•	•	•
CA	•		•	•	•	•	•	•		•
CO	•		•	•	•	•	•	•		
CT		•								
DE	•		•	•		•				
FL	•		•	•	•	•			•	•
GA	•		•	•		•				
HI	•					•	•			•
IA		•								
ID	•		•	•	•	•				
IL	•		•	•		•				
IN	•				•	•				
KS	•		•	•	•	•				
KY		•								
LA		•								
MA		•								
MD	•			•		•				
ME		•								
MI	•				•	•				
MN		•								
MO	•			•		•				
MS		•								

Table 3-3. Summary of State Water Reuse Regulations and Guidelines (cont.)

State	Regulations or Guidelines	No Regulations or Guidelines	Urban Reuse		Agricultural Reuse		Recreational Reuse		Environ-mental Reuse	Indus-trial Reuse
			Unrestricted	Restricted	Food Crops	Nonfood Crops	Unrestricted	Restricted		
MT	•		•	•	•	•				
NC	•			•						
ND	•					•				
NE				•		•				
NH		•								
NJ	•					•				
NM	•		•	•	•	•				
NV	•		•	•	•	•	•	•		•
NY	•					•				
OH		•								
OK	•			•		•				
OR	•		•	•	•	•	•	•		•
PA		•								
RI		•								
SC	•		•	•		•				
SD	•		•	•		•			•	
TN	•		•	•		•				
TX	•		•	•	•	•		•		•
UT	•		•	•	•	•				•
VA		•								
VT	•					•				
WA	•		•	•	•	•				
WI	•					•				
WV	•				•	•				
WY	•		•	•	•	•				

The following subsections present criteria information for six states: California, Florida, Arizona, Nevada, Texas, and Washington. A seventh subsection compares the criteria.

3.4.1 California Criteria

Water quality requirements for reclaimed water systems in the State of California are generally driven by two different motivations: regulatory compliance and customer standards. Monitoring programs for regulatory compliance are primarily based on National Pollution Discharge Elimination System (NPDES) and California Code of Regulations (CCR) requirements. Distribution system monitoring has evolved around internal standards for each of the different types of uses.

For more than 30 years, the State of California Department of Health Services (DOHS) has provided standards in the CCR for the use of reclaimed water for non-potable purposes. The evolution of these standards since 1968 has widened the spectrum of acceptable uses of reclaimed water.

The May 1968 Water Reclamation Standards were covered in the CCR, Title 17–Public Health. These basic regulations covered specific definitions and categories of uses. The 1968 Title 17 included definitions and regulations for irrigation of food crops; irrigation of fodder, fiber and seed crops; landscape irrigation; and recreational impoundments.

In 1978, the CCR Title 22, Division 4–Environmental Health was adopted, which established new standards for water reclamation in California. The 1978 Title 22 regulations included new usage areas and a performance and design section. Components included groundwater recharge, other methods of treatment, sampling and analysis, engineering report and operational requirements, general requirements of design, alternative reliability requirements for uses of permitting primary effluent, and alternative reliability requirements for uses requiring oxidized, disinfected wastewater or oxidized, clarified, filtered, disinfected wastewater.

The California Water Code has revised the terminology for reclaimed water. Reclaimed water is no longer the acceptable terminology; the correct term is now "recycled water." The State DOHS has developed revisions to Title 22 Standards for Water Recycling entitled "Water Recycling Criteria." The new regulations, which became effective on December 2, 2000, have incorporated the following changes from the 1978 version.

- Three levels of wastewater treatment are defined: Disinfected Secondary-2.2 Reclaimed Water, Disinfected Secondary-23 Reclaimed Water and Disinfected Tertiary Reclaimed Water.
- The list of acceptable reclaimed water uses for each level of wastewater treatment is expanded.
- Monitoring requirements are expanded for using reclaimed water in unrestricted recreational impoundments.
- Standards are set for the use of reclaimed water in commercial cooling towers.
- More explicit definitions for use area requirements are established.

- Standards are set for the use of reclaimed water for dual plumbed facilities.
- Language changes for engineering reports and design/operation/and sampling requirements are implemented.

The approved uses, total coliform limits, and treatment levels under the new Title 22 regulations are outlined in Table 3-4.

3.4.2 Florida Criteria

There has been a significant increase in reuse in the last 10 years, and reuse has been an important component of water resource and wastewater management in Florida (http://www.state.fl.us/water/reuse/index.htm). Florida requires the state's water management districts to identify areas where water supplies are critical or are anticipated to be critical. Water reuse feasibility studies for wastewater treatment plants are then required for such areas. State legislation requires a reasonable amount of reclaimed water use unless the reuse is not feasible for technical, environmental, or economic reasons. Florida's water quality criteria, summarized in Table 3-5, are generally more restrictive than the recommendations made by U.S. EPA (1992).

3.4.3 Arizona Criteria

The Arizona Department of Environmental Quality (ADEQ) has recently revised its reuse regulations. The revisions became effective in January 2001. ADEQ established five reclaimed water quality categories that specify minimum levels of treatment. It also adopted a limited number of numeric reclaimed water quality criteria. Wastewater treatment plants are required to monitor for compliance with reclaimed water quality standards either in the plant (for example, turbidity) or at a point where reclaimed water leaves the wastewater treatment plant and enters a canal or a pipeline distribution system. Monitoring requirements are to be established in individual or general permits for the wastewater treatment plant to provide timely data on whether required treatment processes are operating correctly. For example, ADEQ established reclaimed water quality criteria for turbidity, total nitrogen, and fecal coliform bacteria. Compliance with required secondary treatment requirements is to be determined by monitoring compliance with certain conventional pollutants such as pH, TSS, and 5-day BOD (BOD_5). BOD_5 is the total amount of oxygen consumed by microorganism during the first 5 days of biodegradation. ADEQ's revisions of the reclaimed water quality standards were based, in part, on other states' approaches to wastewater reuse regulation, particularly those of California and Florida.

Table 3-4. California Water Recycling Criteria: Treatment and Quality Criteria for Nonpotable Uses of Reclaimed Water

Type of Use	Total Coliform Limits[a]	Treatment Required
Irrigation of fodder, fiber, & seed crops, orchards[b] and vineyards[b], and processed food crops; flushing sanitary sewers	▪ None required	▪ Secondary
Irrigation of pasture for milking animals, landscape areas[c], ornamental nursery stock, and sod farms; landscape impoundments; industrial or commercial cooling water where no mist is created; nonstructural fire fighting; industrial boiler feed; soil compaction; dust control; cleaning roads, sidewalks, and outdoor areas	▪ ≤23/100 mL ▪ ≤240/100 mL in more than one sample in any 30-day period	▪ Secondary ▪ Disinfection
Irrigation of food crops[b]; restricted landscape impoundments; and fish hatcheries	▪ ≤2.2/100 mL ▪ ≤23/100 mL in more than one sample in any 30-day period	▪ Secondary ▪ Disinfection
Irrigation of food crops[d] and open access landscape areas[e]; toilet and urinal flushing; industrial process water; decorative fountains; commercial laundries; snow-making; structural fire fighting; and industrial or commercial cooling where mist is created	▪ ≤2.2/100 mL ▪ ≤23/100 mL in more than one sample in any 30-day period ▪ 240/100 mL (maximum)	▪ Secondary ▪ Coagulation[f] ▪ Filtration[g,h] ▪ Disinfection
Nonrestricted recreational impoundments	▪ ≤2.2/100 mL ▪ ≤23/100 mL in more than one sample in any 30-day period ▪ 240/100 mL (maximum)	▪ Secondary ▪ Coagulation[f] ▪ Clarification[i] ▪ Filtration[g,h] ▪ Disinfection

[a] Based on running seven-day median.
[b] No contact between reclaimed water and edible portion of crop.
[c] Cemeteries, freeway landscaping, restricted access golf courses, and other controlled access irrigation areas.
[d] Contact between reclaimed water and edible portion of crop; includes edible root crops.
[e] Parks, playgrounds, schoolyards, residential landscaping, unrestricted access golf courses, and other uncontrolled access irrigation areas.
[f] Not required if the turbidity of the influent to the filters does not exceed 5 NTU more than 5% of the time within any 24-hour period.
[g] The turbidity after filtration through filter media cannot exceed 2 nephelometric turbidity units (NTU) within any 24-hour period, 5 NTU more than 5% of the time within a 24-hour period, and 10 NTU at any time. The turbidity after filtration through a membrane process cannot exceed 0.2 NTU more than 5% of the time within any 24-hour period and 0.5 NTU at any time.
[h] The filtration rate cannot exceed 5 gpm/ft^2 for conventional media filters or 2 gpm/ft^2 for traveling bridge automatic backwash filters.
[i] Clarification not required if reclaimed water monitoring requirements for enteric viruses, *Giardia*, and *Cryptosporidium* are met.

Table 3-5. Florida Treatment and Water Quality Criteria for Reclaimed Water

Type of Use	Water Quality Limits	Treatment Required
Restricted public access area irrigation[a], industrial uses[b]	200 fecal coli/100 mL 20 mg/L TSS[g] 20 mg/L CBOD	Secondary and disinfection
Public access area irrigation[c], food crop irrigation[d], toilet flushing[e], fire protection, aesthetic purposes, dust control, commercial laundries, vehicle washing, other uses[f]	No detectable fecal coli/100 mL[h] 5.0 mg/L TSS 20 mg/L CBOD	Secondary, filtration, and disinfection
Rapid infiltration basins, absorption fields	200 fecal coli/100 mL 20 mg/L TSS 20 mg/L CBOD 12 mg/L No_3 (as N)	Secondary and disinfection
Rapid infiltration basins in unfavorable geohydrologic conditions	No detectable fecal coli/100 mL[h] 5.0 mg/L TSS Primary and secondary drinking water standards	Secondary, filtration, and disinfection
Injection to groundwater	No detectable fecal coli/100 mL[h] 5.0 mg/L TSS Primary and secondary drinking water standards	Secondary, filtration, and disinfection
Injection to formations of Floridan or Biscayne Aquifers having TDS <500 mg/L	No detectable fecal coli/100 mL[h] 5.0 mg/L TSS 3 mg/L TOC 0.2 mg/L TOX[I] Primary and secondary drinking water standards	Secondary, filtration, disinfection, and activated carbon adsorption
Discharge to Class I surface waters (used for potable supply)	No detectable fecal coli/100 mL[h] 5 mg/L TSS 20 mg/L CBOD 10 mg/L NO_3 (as N) Primary and secondary drinking water standards	Secondary, filtration, and disinfection

[a] Sod farms, forests, pastureland, areas used to grow trees and fodder, fiber, and seed crops, or similar areas.
[b] Contact between reclaimed water and food or beverage products prohibited.
[c] Residential lawns, golf courses, cemeteries, parks, landscaped areas, highway medians, or similar areas.
[d] Direct contact between reclaimed water and tobacco and citrus is allowed, as is direct contact between reclaimed water and edible crops that are peeled, skinned, cooked, or thermally processed before consumption; also allowed for all edible crops where irrigation methods preclude direct contact between reclaimed water and crops.
[e] Only allowed where residents do not have access to plumbing system. Not allowed in single-family residences.
[f] Flushing of sanitary sewers and reclaimed water lines, mixing of cement, manufacture of ice for ice rinks, and cleaning roads, sidewalks, and outdoor areas.
[g] TSS in reclaimed water used for subsurface irrigation systems cannot exceed 10 mg/L.
[h] No detectable fecal coliform organisms/100 mL in at least 75% of the samples, with no single sample to exceed 25 fecal coliform organisms/100 mL.
[i] TOX = total organic halogen.

The five classes of reclaimed water are A+, A, B+, B, and C. Class A+ reclaimed water is wastewater that has undergone secondary treatment, filtration, nitrogen removal treatment, and disinfection, while Class A is wastewater that has undergone secondary treatment, filtration, and disinfection. Class B+ is wastewater that has undergone secondary treatment, nitrogen removal treatment, and disinfection, while Class B is wastewater that has undergone secondary treatment and disinfection. Class C is wastewater that has undergone secondary treatment in a series of wastewater stabilization ponds, including aeration, with or without disinfection. The total retention time of Class C reclaimed water in the stabilization ponds must be at least 20 days.

The reclaimed water quality criteria for turbidity, total nitrogen, and fecal coliform bacteria are presented in Table 3-6. The minimum reclaimed water quality requirements by type of reuse are presented in Table 3-7.

Table 3-6. Arizona Reclaimed Water Quality Criteria

Class	Turbidity	Total Nitrogen	Fecal Coliform
A+	24 hour avg. < 2 NTU, < 5 NTU at all times	< 10 mg/L	None detected in 4 of last 7 daily samples, < 23/100 mL at all times
A	24 hour avg. < 2 NTU, < 5 NTU at all times	N/A	None detected in 4 of last 7 daily samples, < 23/100 mL at all times
B+	N/A	< 10 mg/L	< 200/100 mL in 4 of last 7 daily samples, < 800/100 mL at all times
B	N/A	N/A	< 200/100 mL in 4 of last 7 daily samples, < 800/100 mL at all times
C	N/A	N/A	< 1000/100 mL in 4 of last 7 daily samples, < 4000/100 mL at all times

3.4.4 Nevada Criteria

Nevada issued requirements for the recycling of wastewater on September 13, 1991 under the Nevada Administrative Code, Chapter 445A. The use of treated effluent for irrigation was determined for four categories of uses, all of which require at least secondary treatment. Category A(1) is for irrigation of pastures, while Category A is for irrigation for other agricultural purposes except for growing crops for human consumption where public access is prohibited. Category B is for irrigation of a golf course, cemetery, or greenbelt where public access is controlled and human contact with the treated effluent does not occur; or an impoundment where all activities are prohibited and human contact with the treated effluent does not occur. Category C is for irrigation of a cemetery, highway median, park, playground or residential/commercial lawn where public access is controlled and human contact with the treated effluent cannot reasonably be expected; or an impoundment where full body contact with the treated effluent cannot reasonably be expected. The bacteriological water quality requirements and buffer zone requirements for each use are shown in Table 3-8.

3.4.5 Texas Criteria

The Texas Natural Resource Conservation Commission (TNRCC) revised their administrative code for the use of reclaimed water, which became effective on February 12, 1997. The TNRCC set two categories of reclaimed water use, including irrigation or other uses, depending on whether the public may be present during the time when irrigation takes place or other uses where the public may come in contact with the reclaimed water. Type I use is for cases where the public may be present. Examples of Type I uses include the following:

Table 3-7. Arizona Minimum Reclaimed Water Quality Requirements for Direct Reuse

Type of Direct Reuse	Minimum Class of Reclaimed Water Required
Irrigation of food crops	A
Recreational impoundments	A
Residential landscape irrigation	A
Schoolground landscape irrigation	A
Open access landscape irrigation	A
Toilet and urinal flushing	A
Fire protection systems	A
Spray irrigation of an orchard or vineyard	A
Commercial closed loop air conditioning systems	A
Vehicle and equipment washing (does not include self-service vehicle washes)	A
Snowmaking	A
Surface irrigation of an orchard or vineyard	B
Golf course irrigation	B
Restricted access landscape irrigation	B
Landscape impoundment	B
Dust control	B
Soil compaction and similar construction activities	B
Pasture for milking animals	B
Livestock watering (dairy animals)	B
Concrete and cement mixing	B
Materials washing and sieving	B
Street cleaning	B
Pasture for non-dairy animals	C
Livestock watering (non-dairy animals)	C
Irrigation of sod farms	C
Irrigation of fiber, seed, forage, and similar crops	C
Silviculture	C

Table 3-8. Nevada Water Quality Requirements for Fecal Coliform and Buffer Zone

Category	A	A(1)	B	C
30 day geometric mean	No limit	200	23	2.2
Maximum daily number	No limit	400	240	23
Minimum buffer zone (feet)	800	400	100	0

- Residential irrigation, including landscape irrigation at individual homes.
- Urban uses, including irrigation of public parks, golf courses with unrestricted public access, schoolyards, or athletic fields.
- Use of reclaimed water for fire protection, either in internal sprinkler systems or external fire hydrants.
- Irrigation of food crops where the applied reclaimed water may have direct contact with the edible part of the crop, unless the food crop undergoes a pasteurization process.
- Irrigation of pastures for milking animals.

- Maintenance of impoundments or natural water bodies where recreational activities, such as wading or fishing, are anticipated even though the water body was not specifically designed for such a use.
- Toilet or urinal flush water.
- Other similar activities where the potential for unintentional human exposure may occur.

Type II use is for cases where the public would not come in contact with the reclaimed water. Examples of Type II uses include:

- Irrigation of sod farms, silviculture, limited access highway rights of way, and other areas where human access is restricted or unlikely to occur.
- Irrigation of food crops where the reclaimed water is not likely to have direct contact with the edible part of the crop, or where the food crop undergoes pasteurization prior to distribution for consumption.
- Irrigation of animal feed crops other than pasture for milking animals.
- Maintenance of impoundments or natural water bodies where direct human contact is not likely.
- Soil compaction or dust control in construction areas where application procedures minimize aerosol drift to public areas.
- Cooling tower makeup water (use for cooling towers which produce significant aerosols adjacent to public access areas may have special requirements).
- Irrigation or other nonpotable uses of reclaimed water at a wastewater treatment facility.

Numerical standards were developed for BOD_5, turbidity, and fecal coliforms for the two use types. In addition, Type II use was divided into two subcategories, depending on whether a pond system was used for treatment. The 30 day average for the parameter concentrations is shown in Table 3-9.

Table 3-9. Texas Reclaimed Water Quality Criteria

Type	BOD_5	$CBOD_5$	Turbidity	Fecal Coliform*	Fecal Coliform**
I	5 mg/L	5 mg/L	3 NTU	20 CFU/100 mL	75 CFU/100 mL
II	20 mg/L	15 mg/L		200 CFU/100mL	800 CFU/100mL
II (with pond)	30 mg/L			200 CFU/100mL	800 CFU/100mL

* Geometric mean
** Single grab sample, not to exceed

3.4.6 Washington Standards

The Washington State Departments of Health and Ecology (DOH and DOE) issued "Water Reclamation and Reuse Standards" in September of 1997 which described the allowable beneficial uses and the required level of treatment.

The standards required automated alarms, redundancy of treatment units, and stringent operator training to meet the criteria that reclaimed water be reliably generated. DOH and DOE established four reclaimed water quality categories that specified minimum treatment levels.

They also adopted numerical limits for total coliform bacteria for each reclaimed water category. Numerical limits, which applied to the monitoring of all reclaimed water categories, were also adopted for BOD (monthly mean of 30 mg/L), TSS (monthly mean of 30 mg/L), turbidity (monthly mean of 2 NTU, not to exceed 5 NTU), and DO ("shall contain dissolved oxygen"). Setback distances were required for distribution and use of reclaimed water, which varied between 0 and 1,000 feet, depending on reclaimed water category and the reuse conditions.

The four classes of reclaimed water are A, B, C, and D. Class A reclaimed water is oxidized, coagulated, filtered, and disinfected, while Classes B, C, and D reclaimed waters are oxidized and disinfected. The reclaimed water quality criteria for total coliforms are presented in Table 3-10. The treatment and quality requirements by type of use are presented in Table 3-11.

Table 3-10. Washington Reclaimed Water Criteria for Coliform

Class	Total Coliform
A	Median < 2.2/100 mL for last 7 days, < 23/100 mL at all times
B	Median < 2.2/100 mL for last 7 days, < 23/100 mL at all times
C	Median < 23/100 mL for last 7 days, < 240/100 mL at all times
D	Median < 240/100 mL for last 7 days

3.4.7 Comparison of State Standards

The project team compared standards for the six above states for various types of irrigation uses. Irrigation is one of the most common uses of reclaimed water, and each state has set water quality limits and/or treatment requirements when allowing the use of reclaimed water. For the irrigation of pastures, with milking animals allowed, the water quality limits and treatment requirements are summarized in Table 3-12. Five of the six states set total or fecal coliform limits, while four set BOD limits. Four of the six states have secondary treatment and disinfection requirements.

Stricter requirements were set for irrigation of restricted access areas and are summarized in Table 3-13. All of the states set total or fecal coliform limits, while three set BOD limits. Four have secondary treatment and disinfection requirements.

For urban irrigation, the water quality limits and treatment requirements are summarized in Table 3-14. All of the states set total or fecal coliform limits, while only three set BOD limits. Four have secondary treatment and disinfection requirements.

For the irrigation of food crops eaten raw, water quality limits and treatment requirements were set by only three of the six states, as shown in Table 3-15. Irrigation of food crops was not allowed in the other three states. Water quality limits are determined by fecal coliform limits, turbidity limits, and/or BOD and TSS requirements. All three states have filtration and disinfection requirements.

Table 3-11. Washington Treatment and Quality Requirements for Reclaimed Water Use

Use	Type of Reclaimed Water Allowed			
	Class A	Class B	Class C	Class D
Irrigation of Nonfood Crops				
Trees and fodder, fiber, and seed crops	Yes	Yes	Yes	Yes
Sod, ornamental plants for commercial use, and pasture to which milking cows or goats have access	Yes	Yes	Yes	No
Irrigation of Food Crops				
Spray irrigation				
All food crops	Yes	No	No	No
Food crops which undergo physical or chemical processing sufficient to destroy all pathogenic agents	Yes	Yes	Yes	Yes
Surface Irrigation				
Food crops where there is no reclaimed water contact with edible portion of crop	Yes	Yes	No	No
Root crops	Yes	No	No	No
Orchards and vineyards	Yes	Yes	Yes	Yes
Food crops which undergo physical or chemical processing sufficient to destroy all pathogenic agents	Yes	Yes	Yes	Yes
Landscape Irrigation				
Restricted access areas (e.g., cemeteries and freeway landscapes)	Yes	Yes	Yes	No
Open access areas (e.g., golf courses, parks, playgrounds, schoolyards, and residential landscapes)	Yes	No	No	No
Impoundments				
Landscape impoundments	Yes	Yes	Yes	No
Restricted recreational impoundments	Yes	Yes	No	No
Nonrestricted recreational impoundments	Yes	No	No	No
Fish Hatchery Basins	Yes	Yes	No	No
Decorative Fountains	Yes	No	No	No
Flushing of Sanitary Sewers	Yes	Yes	Yes	Yes
Street Cleaning				
Street sweeping, brush dampening	Yes	Yes	Yes	No
Street washing, spray	Yes	No	No	No
Washing of Corporation Yards, Lots, and Sidewalks	Yes	Yes	No	No
Dust Control (dampening unpaved roads and other surfaces)	Yes	Yes	Yes	No
Dampening of Soil for Compaction (at construction sites, landfills, etc.)	Yes	Yes	Yes	No
Water Jetting for Consolidation of Backfill around Pipelines				
Pipelines for reclaimed water, sewage, storm drainage, and gas; and conduits for electricity	Yes	Yes	Yes	No
Fire Fighting and Protection				
Dumping from aircraft	Yes	Yes	Yes	No
Hydrants or sprinkler systems in buildings	Yes	No	No	No
Toilet and Urinal Flushing	Yes	No	No	No
Ship Ballast	Yes	Yes	Yes	No
Washing Aggregate and Making Concrete	Yes	Yes	Yes	No
Industrial Boiler Feed	Yes	Yes	Yes	No
Industrial Cooling				
Aerosols or other mist not created	Yes	Yes	Yes	No
Aerosols or other mist created (e.g., use in cooling towers, forced air evaporation, or spraying)	Yes	No	No	No
Industrial Process				
Without exposure of workers	Yes	Yes	Yes	No
With exposure of workers	Yes	No	No	No

Table 3-12. Standards for Pasture Irrigation with Reclaimed Water in AZ, CA, FL, NV, TX, and WA (milking animals allowed)

State	Water Quality Limits	Treatment Required
Arizona	126 fecal coliform/100 mL 30 mg/L BOD 30 mg/L TSS	Secondary Disinfection
California	23 total coliform/100 mL	Secondary Disinfection
Florida	200 fecal coliform/100 mL 20 mg/L BOD 20 mg/L TSS	Secondary Disinfection
Nevada	No fecal coliform limit with 800 ft buffer zone; 400 fecal coliform/100 mL max., 200 fecal coliform/100 mL avg. with 400 ft buffer zone	Secondary Disinfection
Texas	30 mg/L BOD	Not Specified
Washington	23 total coliform/100 mL 30 mg/L BOD 30 mg/L TSS	Oxidation Disinfection

Table 3-13. Standards for Irrigation for Restricted Access Areas with Reclaimed Water in AZ, CA, FL, NV, TX, and WA

State	Water Quality Limits	Treatment Required
Arizona	126 E. coli/100 mL 30 mg/L BOD 30 mg/L TSS	Secondary Disinfection
California	23 total coliform/100 mL	Secondary Disinfection
Florida	200 fecal coliform/100mL 20 mg/L BOD 20 mg/L TSS	Secondary Disinfection
Nevada	23 fecal coliform/100mL max., 240 fecal coliform/100ml avg. with 100 ft buffer zone	Secondary Disinfection
Texas	800 fecal coliform/100mL 20 mg/L TSS	None Specified
Washington	23 total coliform/100 mL 30 mg/L BOD 30 mg/L TSS	Oxidation Disinfection

Table 3-14. Standards for Urban Irrigation with Reclaimed Water in AZ, CA, FL, NV, TX, and WA

State	Water Quality Limits	Treatment Required
Arizona	No detectable fecal coliform/100mL 2 NTU	Secondary Filtration Disinfection
California	2.2 total coliform/100mL 2 NTU	Secondary Filtration Disinfection
Florida	No detectable fecal coliform/100mL 20 mg/L BOD 5 mg/L TSS	Secondary Filtration Disinfection
Nevada	2.2 fecal coliform/100mL max., 23 fecal coliform/100mL avg.	Secondary Disinfection
Texas	75 fecal coliform/100 mL 3 NTU 5 mg/L BOD	None Specified
Washington	2.2 total coliform/100 mL 30 mg/L BOD 30 mg/L TSS	Oxidation Coagulation Filtration Disinfection

Table 3-15. Standards for Reclaimed Water Spray Irrigation of Food Crops Eaten Raw in AZ, CA, FL, NV, TX, and WA

State	Water Quality Limit	Treatment Required
Arizona	No detectable fecal coliform/100 mL 2 NTU No detectable parasites	Secondary Filtration Disinfection
California	2.2 total coliform/100 mL 2 NTU	Secondary Filtration Disinfection
Florida	Use prohibited	
Nevada	Use prohibited	
Texas	Use prohibited	
Washington	2.2 total coliform/100 mL 30 mg/L BOD 30 mg/L TSS	Oxidation Coagulation Filtration Disinfection

CHAPTER 4.0

WATER QUALITY PARAMETERS

4.1 Overview

As discussed in Chapter 3.0, treated wastewater can be reused for a wide variety of applications, and regulatory guidelines and standards vary, depending on the intended reuse. Reuse projects are often seasonal in nature and require storage of reclaimed water for long periods. Frequently, the water is held in open storage reservoirs. During periods of storage, the quality of reclaimed water can change. Water quality issues associated with storage of reclaimed water in open reservoirs are reviewed in this chapter. A case study of the Sand Canyon Reservoir and the Rattlesnake Reservoir in Southern California is also provided.

Many of the significant water quality parameters that impact the beneficial uses of open reclaimed (or nonpotable) water reservoirs are the same as for freshwater reservoirs. These include physical, chemical, biological, and aesthetic water quality parameters. Perhaps the greatest difference is that the water supply for reclaimed (or nonpotable) water reservoirs can have significantly higher levels of nutrients (phosphorous and nitrogen). Differences in nutrient levels can have a direct effect on biological uptake and alga responses in reclaimed water reservoirs.

For the case study provided, Sand Canyon Reservoir is predominantly made up of reclaimed water with local runoff. Reclaimed water delivered from the Michelson Water Reclamation Plant (MWRP) to Sand Canyon Reservoir typically makes up over 90% of the water in storage. Rattlesnake Reservoir is made up of untreated surface water from Irvine Lake via the Irvine Lake Pipeline (ILP) and reclaimed water from the MWRP, and varies dramatically with demands as it is located the farthest distance from the MWRP.

Reclaimed (or nonpotable) water reservoirs are typically smaller than freshwater reservoirs with sizes commonly below 1,000 acre-feet (86,000 m^3) in storage capacity. These reservoirs are used primarily for operational purposes, as opposed to flood control storage that occurs during storm events. During the off peak demand periods (fall and winter), the reservoirs are filled, while maintaining a certain freeboard for runoff. However, during the summer months, when reclaimed water use is the greatest, they are used to manage daily fluctuations in demand.

Reclaimed water reservoirs follow the same traditional thermal patterns within the water column of stratification/destratification as freshwater reservoirs, resulting in nutrient cycling and algae growth responses. However, due to excessive levels of nutrients availability, traditional biological responses and relationships found in freshwater systems cannot be used. Some of these responses include determining the limiting nutrient using nitrogen/phosphorous ratios and the algae responses based on phosphorous and chlorophyll *a*.

Additional water quality parameters that impact the potential use of reclaimed water in storage reservoirs are of an aesthetic nature. Some of these include hydrogen sulfide (H_2S) and color, which can impact specialty users such as high rise toilet flushing and carpet dyeing plants. General inorganic quality [alkalinity, total dissolved solids (TDS), hardness] can also have an impact upon the potential usage of reclaimed water by specialty applications as well as traditional usage (green belt irrigation, agricultural, turf grass).

4.2 Physical Properties

The physical properties of reclaimed water reservoirs that affect water quality are considered in the following discussion. Most of the literature on the relationships of physical properties to water quality is based upon freshwater lake and reservoir research. Physical properties are discussed in this context, and an actual case study for two reclaimed water storage reservoirs is included to illustrate these concepts.

Physical properties having an impact upon a reservoir's water quality and subsequent discharge quality include surface area, storage volume, and maximum and mean depths, which in turn determine the residence times. Reclaimed water reservoirs are usually shallow, with maximum water depths less than 33' (10 meters), which is typical of operational reclaimed water reservoirs and small lakes used in golf courses for irrigation purposes.

Water quality is also impacted by water inputs (reclaimed water, precipitation, and runoff) and water withdrawal activities. Watershed factors such as land use characteristics (level of development) ultimately determine the amount of nutrient and suspended solids loadings to the reservoir.

Temperature and dissolved oxygen (DO) levels in the reservoir play important roles in determining the water quality in a reclaimed water reservoir. For this reason, the primary discussion in this section focuses on temperature and DO, although pH is also discussed.

4.2.1 Temperature

Temperature gradients within the water column along with wind action at the surface have a direct impact on mixing in reservoirs and lakes. Density variations must also be taken into account in determining the potential for mixing within a reservoir. Wetzel (1983) found that the amount of energy required to mix two water masses of different temperatures is proportional to their density differences, which is why water masses of different temperatures and thus different densities are resistant to mixing.

Because the majority of reclaimed water storage reservoirs are typically shallow, the potential consequences on water quality must be taken into consideration. Studies by Allen and Kramer (1972) for freshwater reservoirs have shown that wind patterns at the lake surface and currents within the water column play a significant role in the partial destratification of shallow reservoirs. They noted that this change in stratification pattern brought about by maintaining a shallow reservoir increases the amount of vertical mixing of the upper and lower waters, which enhances the internal transport of nutrients such as iron, manganese, and phosphorous from the lake sediments to the water column.

Lakes and reservoirs gain and lose energy through the surface because of shear wind forces, heating from the sun, and radiant cooling. As the weather warms, the water column gains heat, but remains uniform in temperature, or isothermal. Without density differences, resistance to mixing is reduced. With the onset of spring, surface waters warm rapidly, become lighter, and cannot mix with the denser and colder bottom water layer. From spring through summer, this warm upper layer of water, called the epilimnion, circulates, mixes, and exchanges gases with the atmosphere. In contrast to the epilimnion, the lower layer of colder water, or hypolimnion, is weakly mixed and usually has distinct gradients in nutrient and oxygen concentrations near the bottom. Between the epilimnion and hypolymnion is the metalimnion or thermocline, a thin layer within the water column having a sharp temperature gradient (Cooke and Carlson, 1989).

In the fall, heat is lost to the atmosphere, and surface waters become cooler and more dense. This occurrence causes the surface waters to sink to a level of equivalent density. The cooling and sinking of the water, combined with fall winds, break down the thermal stratification and result in isothermal conditions once again. Fall mixing is often called turnover (Hutchinson, 1957).

Cooke and Carlson (1989), Tchobanoglous and Schroeder (1985), and Mitchell (1972) have recognized that thermal stratification produces two layers with significant differences in water quality. The warmer, well lighted epilimnion will be high in DO, may support algae, and typically will have lower concentrations of elements such as iron, manganese, and phosphorous. In contrast, the hypolymnion may be stagnant and is isolated from gas exchange with the air. It is colder and dark, with little or no oxygen-generating photosynthesis occurring. If the reservoir receives high organic loading and/or supports algae growth, the deeper waters will be rich in organic matter, and subsequent consumption of oxygen by microbes will probably exceed oxygen generation by deep water algae, resulting in the depletion of oxygen within the hypolymnion.

The effects of organic loadings and extensive algal growth and decay can lead to increased color levels caused by algae blooms and potentially to the production of odors (H_2S). H_2S production occurs in the hypolimnion, where oxygen levels are low under reducing conditions. These aesthetic parameters can have a dramatic impact upon the use of reclaimed water in operational reservoirs.

4.2.2 Dissolved Oxygen

Dissolved oxygen in a reservoir is derived from two sources: from the atmosphere and from photosynthetic inputs (algae). The distribution of oxygen in the water column can be determined by the rate of consumption by bacteria and other organisms and/or photosynthetic input by algae and atmospheric oxygen transfer.

The solubility of many inorganic nutrients is affected by the resulting oxygen distribution in the water column. Seasonal shifts from an aerobic to an anaerobic environment, which occur in the summer months, provide the nutrients available for rapid growth of many organisms. Nutrients affected by the oxygen distribution include phosphorous, iron, and manganese.

4.2.3 Hydrogen Ion Activity (pH)

pH is a measure of the hydrogen ion activity in a water sample. It is mathematically related to the hydrogen ion activity according to the expression:

$$pH = -\log \{H^+\}, \text{ where } \{H^+\} \text{ is the hydrogen ion activity.}$$

The pH of natural waters is a measure of acid-base equilibrium achieved by the various dissolved compounds, salts, and gases. The carbonate system in natural waters is primarily responsible for regulating pH and is composed of carbon dioxide (CO_2), carbonic acid (H_2CO_3), bicarbonate ion (HCO_3^-), and carbonate ion (CO_3^{-2}). Stumm and Morgan (1970) have described the interactions and kinetics of the carbonate system in some detail.

The biological processes occurring in the lake influence the pH of the water. Photosynthesis near the surface causes the pH to rise. Respiration below the thermocline causes the pH to fall. Under isothermal conditions, the pH is almost constant from top to bottom. Under stratified conditions, the pH at the bottom is much lower than at the surface. An excessive production of oxygen at the surface of the lake, (photosynthesis >> than respiration) is paralleled by a large decrease in oxygen, or anaerobic conditions at the bottom of the lake (respiration >> photosynthesis).

Typically, lower pH values enhance the dissolution of phosphorous, iron, and manganese from the lower stratum and the sediments. Laboratory equilibrium studies conducted by Delfino and Lee (1971) under anaerobic conditions indicated that pH was an important parameter involved with the release of manganese from bottom sediments. These minerals are available as nutrients for continued phytoplankton growth.

4.3 Nutrients

4.3.1 Reclaimed Water

The primary nutrients used by algae are nitrogen and phosphorous. Nitrogen and phosphorous levels in traditional secondary treated and reclaimed waters are usually found at levels sufficient to promote algae growth in reservoirs, if not controlled properly.

The water quality within a reservoir used to store reclaimed water is partly determined by the level of treatment of the source water delivered along with runoff and internal nutrient cycling. Typical reclaimed water effluent nutrient levels for a traditional activated sludge plant with and without nitrogen removal are compared in Table 4-1.

Table 4-1 Comparison of Reclaimed Water Effluent with and without Nitrogen Removal[3]

Constituent	Denitrification/Nitrification[1]		Activated Sludge (no nutrient removal)[2]	
	Average	Standard Deviation (SD)	Average	Standard Deviation (SD)
Total Suspended Solids (TSS), mg/L	1.5	0.7	1.9	1.1
BOD_5, mg/L	<5	1.4	<5	2.0
Turbidity, NTU	<2	0.5	<2	0.5
Chemical Oxygen Demand (COD), mg/L	23	4.1	25	5.3
Total Inorganic Nitrogen as N(TIN), mg/L	5.2	1.4	10.8	4.8
Total Coliform, MPN	<2.2	0.8	<2.2	0.8
Total Phosphorous as P, mg/L	1.78	0.93	1.87	0.66

IRWD Laboratory data for Michelson Water Reclamation Plant: (1) 1999 annual average, (2) 1997-98 annual average, (3) IRWD uses modified Ludzack-Leutinger nutrient removal process for nitrogen removal.

The levels of nitrogen and phosphorous for the reclaimed water are an order of magnitude higher than those found in freshwater lakes or reservoirs. There is an abundance of these essential nutrients to promote algae growth in a reservoir. The additional treatment brought about by the denitrification/nitrification process can result in lower nutrient levels in the plant effluent and ultimately the reclaimed water reservoir. The level of treatment on the reservoir source water will have a direct impact upon the water quality within the reservoir.

A previous study was conducted in December 1993 on secondary effluent at the MWRP. Ortho-phosphate reduction demonstrated with microfiltration (MF) alone removed 50-55% with ferric chloride addition at an optimum dosage of 20 mg/L providing an additional 30-40% reduction. Similar results were found for alum, but, required significantly higher dosages (Spangenberg and Kalinsky, 1993).

4.3.2 Reservoir

The quantity of biomass or algae in a lake or reservoir, an indicator of the degree of eutrophication, is usually limited by the concentration of an essential element or nutrient, such as phosphorous or nitrogen. Phosphorous in the form of dissolved phosphate is considered to be the most important form of phosphorous, as algae can readily uptake phosphorous in this form for growth. Sources of phosphorous include the water column, runoff, the source water, and sediments. Details on the importance of phosphorous and the interaction between the water column and sediment-water exchange are discussed in Section 4.4.3 Sediments, due to the importance of phosphorous exchange at these two continuums.

Several forms of nitrogen can be found in fresh waters: dissolved molecular nitrogen (N_2), a large number of organic compounds which are part of plants and microorganisms, refractory organic compounds, ammonium (NH_4^+), nitrite (NO_2^-), and nitrate (NO_3^-). Sources of nitrogen include: (1) precipitation at the lake surface, (2) nitrogen fixation in the water and sediments, and (3) inputs from surface water drainage. Losses of nitrogen can occur at the effluent from the reservoir, reduction of NO_3^- to N_2 by bacterial denitrification with subsequent return of N_2 to the atmosphere, and permanent loss of inorganic and organic nitrogen compounds to the sediments. The interrelationships between these various forms of nitrogen and the nitrogen cycle in the aquatic and soil environment are illustrated in Figure 4-1.

Studies by Wetzel (1975) have shown that ammonia (NH_3) levels of well-oxidized waters are usually low. Where appreciable amounts of organic matter reach the hypolymnion in stratified lakes, NH_3 tends to accumulate. Under aerobic conditions, bacteria convert NH_3 to NO_2^- and then to NO_3^-, a process known as nitrification. With the loss of the oxidized zone at the sediment/water interface under reduced hypolimnetic conditions, the adsorption capacity of the sediments is also reduced. This results in an increased release of NH_4^+ from the sediments.

Nitrogen to phosphorous ratios are typically used by limnologists throughout the world as an indication of whether water is phosphorous- or nitrogen-limited.

According to Smith (1979), a total nitrogen to total phosphorous (N:P) ratio in excess of 21 strongly supports the hypothesis that phosphorous is the limiting element in supporting algal growth. However, when the ratio of (N:P) is less than 13, nitrogen may be the limiting factor. Ratios found between 13 and 21 could be limited from time to time by either element.

It is important to note that these ratios were developed for natural systems and may not apply to reclaimed water reservoirs. In studies conducted by Goldman and Horne (1983), it was found that phosphorous levels in reclaimed water effluents are at least 5 to 10 times higher than the amount needed to control algae growth in natural systems. Total inorganic nitrogen (TIN) levels are typically above 0.1 milligrams per liter (mg/L) for reclaimed waters, indicating that nitrogen may not limit the growth of phytoplankton.

4.4 Biological Characteristics

Biological constituents play an important role in defining the water quality or health of a reservoir. Most of the constituents measured such as chlorophyll and phytoplankton identification and numeration are a measure of the responses within the reservoir and are primarily determined by varying levels of nutrient inputs and the associated chemical, biological, and physical transformations within the reservoir. The significance of chlorophyll, phytoplankton, and sediment nutrient exchange are summarized in this section followed by a review of key fate processes and transformations.

4.4.1 Chlorophyll

Chlorophyll is a common indicator of the phytoplankton biomass within freshwater lakes. Blue green algae and green algae are the predominant phytoplankton commonly responsible for a major portion of measured chlorophyll values. Chlorophyll is a photosynthetic pigment in all plants that converts sunlight energy into useful energy for plant growth and reproduction. The process of photosynthesis is illustrated by the general equation:

$$CO_2 + PO_4 + NH_3 + H_2O \text{ Yields New Cell Growth} + O_2 + H_2O$$

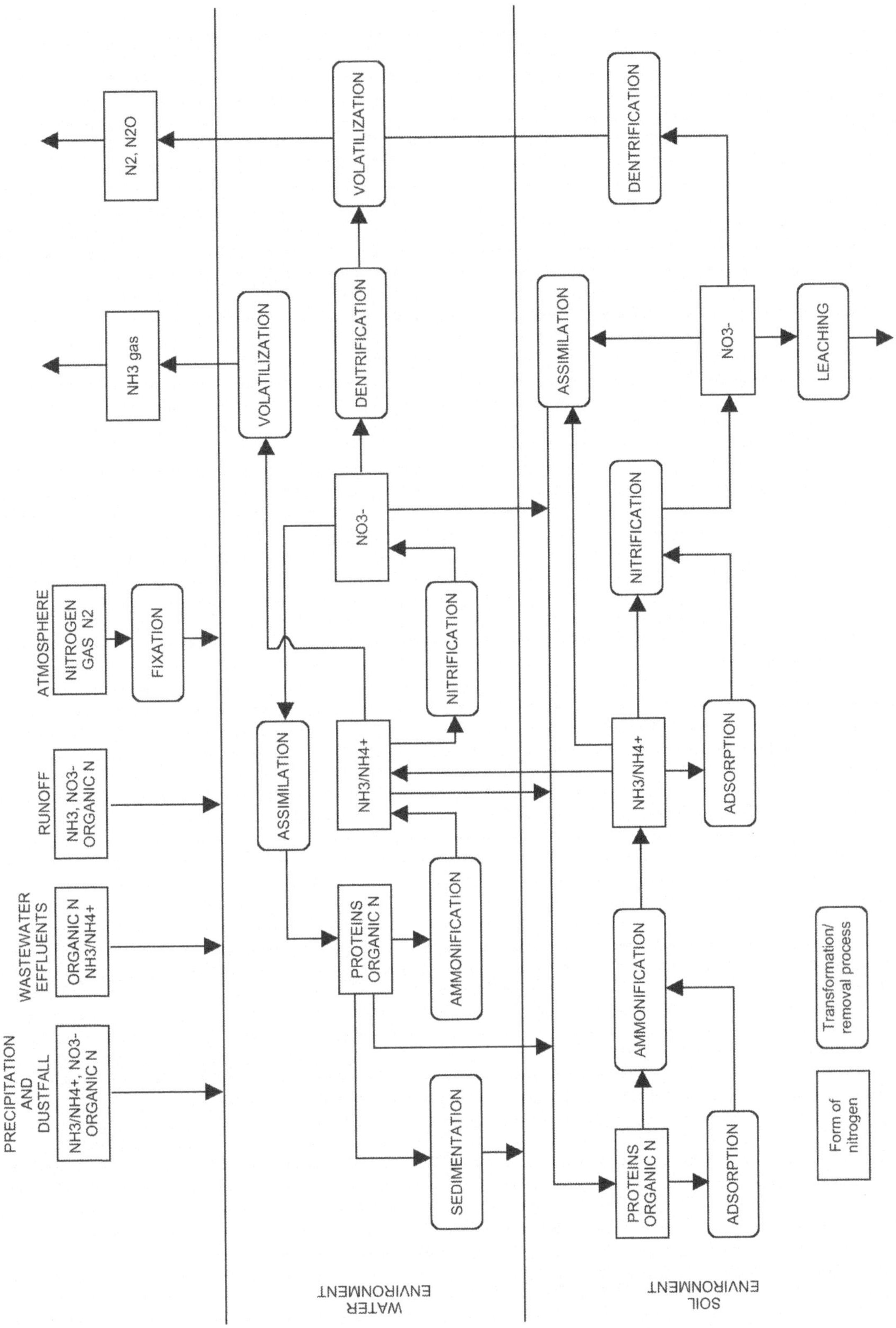

Figure 4-1. Nitrogen Cycle in the Aquatic and Soil Environment

The principal type of chlorophyll found in fresh water reservoirs is chlorophyll *a* and is found in all oxygen evolving organisms. Higher plants and green algae contain chlorophyll *b*. Chlorophyll *c* is present in diatoms and brown algae. Two other types of chlorophyll, chlorophyll *d* and chlorophyll *e* have also been identified, but to a much lesser extent than chlorophyll *a*, *b*, and *c* (Parker, 1984).

The euphotic zone constitutes the upper layers of a reservoir in which sufficient light penetrates to permit the growth of green plants. Studies conducted by limnologists Uhlman and Klapper (1985) and Bernhardt and Clasen (1985) determined that, for most reservoirs, the euphotic zone is located within the upper 10' (3 meters) of the water column.

4.4.2 Phytoplankton

Studies by Chu (1943) and Rodhe (1948) categorized freshwater algae according to tolerance ranges for optimum growth based on phosphorous levels. Three categories were developed: 1) pigmented flagellates such as dinobryon and uroglena have optimum growth and upper tolerance limits when phosphorous levels are below 20 micrograms (μg) phosphate as phosphorous per liter (PO_4-P/L); 2) diatoms such as synedra and fragilaria optimize growth when levels are below 20 μg PO_4-P/L, but have an upper tolerance level above 20 μg PO_4-P/L; and 3) blue green and green algae such as oscillatoria and scenedesmus have an optimum growth and upper tolerance limit when levels are above 20 μg PO_4-P/L.

4.4.3 Sediments

Sediments are typically reservoirs of nutrients found in the particulate form. Iron, manganese, and phosphorous are nutrients that are quite stable in an aerobic, high pH environment. When the sediment surface becomes anoxic, large quantities of iron, manganese, and phosphorous can be released into the hypolimnion, which in shallow lakes may further accelerate eutrophication. In the anoxic hypolimnion, H_2S and NH_4^+ accumulate, and NO_3^- is consumed by denitrification.

The phosphorous content of a lake depends on 1) external loading, 2) phosphorous export via the effluent, 3) sedimentation of phosphorous, and 4) the release of phosphorous from its sediment.

Mortimer (1941, 1942, and 1971) performed superb experimental work that demonstrated the effectiveness of the oxidized microzone between the sediment/water interface in preventing significant release of phosphorous, iron, and manganese. At DO concentrations greater than 2 mg/L, sediment release of nutrients was minimal; below 2 mg/L, major releases of NH_4^+, phosphorous, iron, and manganese result.

4.5 Fate Processes and Transformations

The focus of the discussion is to provide background information on the key processes for the removal and transformation of various water quality parameters in the aquatic environment. A summary of constituent transformation and removal processes is provided in Table 4-2. In addition, typical rate expressions for selected processes are defined. A summary of the typical rate expressions for the process identified in Table 4-2 is given in Table 4-3. As mentioned previously, the key water quality parameters in a reservoir include nitrogen, phosphorous, DO, H_2S, suspended solids, and algae.

Table 4-2. Constituent Transformation and Removal Processes (i.e., Fate Processes) in the Environment[a]

Process	Comments	Constituents' Effects
Adsorption/ Desorption	Many chemical constituents tend to attach or sorb onto solids. The implication for wastewater discharges is that substantial fractions of some toxic chemicals are associated with the suspended solids in the effluent. Adsorption combined with solids settling results in the removal from the water column of constituents that might not otherwise decay.	Metals, trace organics, NH_4^+, PO_4^{-3}
Algal Synthesis	The synthesis of algal cell tissue using the nutrients found in wastewater.	NH_4^+, NO_3^-, PO_4^{-3}, pH, etc.
Bacterial Conversion	Bacterial conversion (both aerobic and anaerobic) is the most important process in the transformation of constituents released to the environment. The exertion of BOD and NOD are the most common examples of bacterial conversion encountered in water-quality management. The depletion of oxygen in the aerobic conversion of organic wastes is also known as deoxygenation. Solids discharged with treated wastewater are partly organic. Upon settling to the bottom, the solid decompose bacterially either anaerobically or aerobically, depending on local conditions. The bacterial transformation of toxic organic compounds is also of great significance.	BOD_5, nitrification, denitrification, sulfate reduction, anaerobic fermentation (in bottom sediments), conversion of priority organic pollutants, etc.
Chemical Reactions	Important chemical reactions that occur in the environment include hydrolysis, photochemical, and oxidation-reduction reactions. Hydrolysis reactions occur between contaminants and water.	Chemical disinfection, decomposition of organic compounds, specific ion exchange, element substitution
Filtration	Removal of suspended and colloidal solids by straining (mechanical and chance contact), sedimentation, interception, impaction, and adsorption.	Total suspended solids, colloidal particles
Flocculation	Flocculation is the term used to describe the aggregation of smaller particles into larger particles that can be removed by sedimentation and filtration. Flocculation is brought about by Brownian motion, differential velocity gradients, and differential settling in which large particles overtake smaller particles and form larger particles.	Colloidal and small particles
Gas absorption/ Desorption	The process whereby a gas is taken up by a liquid is known as absorption. For example, when the DO concentration in a body of water with a free surface is below the saturation concentration in the water, a net transfer of oxygen occurs from the atmosphere to the water. The rate of transfer (mass per unit time per unit surface area) is proportional to the amount by which the DO is below saturation. The addition of oxygen to water is also known as reaeration. Desorption occurs when the concentration of the gas in the liquid exceeds the saturation value and there is a transfer from the liquid to the atmosphere.	O_2, CO_2, CH_4, NH_3, H_2S

Table 4-2. Constituent Transformation and Removal Processes (i.e., Fate Processes) in the Environment[a] (cont.)

Process	Comments	Constituents' Effects
Natural Decay	In nature, contaminants will decay for a variety of reasons, including mortality in the case of bacteria and photooxidation for certain organic constituents. Natural and radioactive decay usually follow first-order kinetics.	Plants, animals, algae, fungi, protozoa, eubacteria (most bacteria), archaebacteria, viruses, radioactive substances, plant mass
Photochemical Reactions	Solar radiation is known to trigger a number of chemical reactions. Radiation in the near-ultraviolet (UV) and visible range is known to cause the breakdown of a variety of organic compounds.	Oxidation of inorganic and organic compounds
Photosynthesis/ Respiration	During the day, algal cells in water bodies will produce oxygen by means of photosynthesis. DO concentrations as high as 30 to 40 mg/L have been measured. During the evening hours, algal respiration will consume oxygen. Where heavy growths of algae are present, oxygen depletion has been observed during the evening hours.	Algae, duckweed, submerged macrophytes, NH_4^+, PO_4^{-3}, pH, etc.
Sedimentation	The suspended solids discharged with treated wastewater ultimately settle to the bottom of the receiving water body. This settling is enhanced by flocculation and hindered by ambient turbulence. In rivers and coastal areas, turbulence is often sufficient to distribute the suspended solids over the entire water depth.	TSS
Sediment Oxygen Demand	The residual solids discharged with treated wastewater will, in time, settle to the bottom of streams and rivers. Because the particles are partly organic, they can be decomposed anaerobically as well as aerobically, depending on conditions. Algae which settle to the bottom will also be decomposed, but much more slowly. The oxygen consumed in the aerobic decomposition of material in the sediment represents another DO demand in the water body.	O_2, particulate BOD
Volatilization	Volatilization is the process whereby liquids and solids vaporize and escape to the atmosphere. Organic compounds that readily volatilize are known as VOCs (volatile organic compounds). The physics of this phenomenon are very similar to gas absorption, except that the net flux is out of the water surface.	VOCs, NH_3, CH_4, H_2S, other gases

[a] From Tchobanoglous et al. (2003).

Phosphorous and nitrogen constituents in a reservoir are of primary concern and are transformed via adsorption/desorption, photosynthesis/respiration, and algal synthesis processes. Dissolved oxygen and H_2S levels are determined by gas absorption/desorption, photosynthesis/repiration, bacterial conversion and sediment oxygen demand. Suspended solids and turbidity levels are determined by sedimentation, photosynthesis/respiration, filtration, and flocculation processes.

4.6 Aesthetic Parameters

Two important aesthetic water quality parameters that can have a direct effect on the potential usage of reclaimed water in storage reservoirs are H_2S and color.

Table 4-3. Typical Rate Expressions for Selected Processes

Process	Rate Expression	Comments
Bacterial Conversion	$r_c = -kC$	r_c = rate of conversion, M/L^3T k = first order reaction rate coefficient, 1/T C = concentration of organic material remaining, M/L^3
Chemical Reactions	$r_c = \pm kC^n$	r_c = rate of conversion, M/L^3T k = reaction rate coefficient, $(M/L^3)^{n-1}$ /T C = concentration of constituent, $(M/L^3)^n$ n = reaction order (e.g. for second order n = 2)
Gas Absorption/ Desorption	$r_{ab} = -k_{ab}\frac{A}{V}(C_s - C)$ $r_{de} = -k_{de}\frac{A}{V}(C - C_s)$	r_{ab} = rate of absorption, M/L^3T r_{de} = rate of desorption, M/L^3T k_{ab} = coefficient of absorption, L/T k_{de} = coefficient of desorption, L/T A = area, L^2 V = volume, L^3 C_s = saturation concentration of constituent in liquid, M/L^3 C = concentration of constituent in liquid, M/L^3
Natural Decay	$r_d = -k_d N$	r_d = rate of decay, no./T k_d = first order reaction rate coefficient, 1/T N = amount of organisms remaining, no.
Sedimentation	$r_s = \frac{v_s}{H}(SS)$	r_s = rate of sedimentation, 1/T v_s = settling velocity, L/T H = depth, L SS= settleable solids, L^3/L^3
Volatilization	$r_v = -k_v (C - C_s)$	r_v = rate of volatilization per unit time per unit volume, M/L^3T k_v = volatilization constant, 1/T C = concentration of constituent in liquid, M/L^3 C_s = saturation concentration of constituent in liquid, M/L^3

[a] From Tchobanoglous et al. (2003), were M = mass, L = length, and T = time.

4.6.1 Hydrogen Sulfide

Sulfate reduction is a bacterial reaction in which bacteria use the oxygen in sulfate to oxidize organic matter to carbon dioxide, producing various sulfide species as a by-product. Stoichiometrically, the reaction is represented by:

$$SO_4^{2-} + 2C\ organic + 2H_2O \rightarrow H_2S + 2HCO_3^-$$

The consequences of sulfate reduction are many. H_2S is fairly toxic to most organisms, particularly fish. The presence of H_2S typically indicates that DO levels are low. Fish kills in reservoirs have been documented when reservoirs and lakes turn over and H_2S and other compounds are mixed into the water column. The sulfide species also impart an odor that can be offensive to customers using reclaimed water.

4.6.2 Color

Color, a visual parameter that, in natural waters, is derived as a result of dissolved materials, is caused primarily by decaying organic matter. Water that has come in contact with the organic matter is usually associated with a light-yellowish to brownish color. Generally, reclaimed water has a light brown tint prior to disinfection. In a reservoir, the sediments act as a sink for decaying plants (primarily algae). As with the production of H_2S, increasing color levels in reservoir discharge waters could be due to lake or reservoir turnover. The consequences of increased color levels are usually minor for irrigation practices, but can have dramatic consequences for specialty applications of reclaimed water such as toilet flushing and carpet dyeing, along with public perception.

4.7 Case Study

Irvine Ranch Water District (IRWD) in Southern California operates two open-storage reclaimed water reservoirs, Rattlesnake and Sand Canyon Reservoirs. The reservoirs were selected for the case study as the water quality within the water column and interactions at the water-sediment interface is generally representative of reservoirs located in the arid, southwest region of the United States where reclaimed water is used extensively. Similar relationships have been documented at two other reservoirs in this region, Santiago Reservoir and Upper Oso Reservoir (Spangenberg, 1991; Jordon, 2001). Some of the dynamics of water quality changes are demonstrated by the temperature and DO level distribution within the water column and generally follow similar patterns in this region of the United States. The behavior of temperature and DO water quality parameters in a reservoir are a function of latitude and altitude. Minor differences in their vertical distribution within the reservoirs located in the middle to eastern portions of the United States are due to extreme seasonal temperature differences (Hutchinson and Loffler, 1956).

Rattlesnake and Sand Canyon Reservoirs receive tertiary treated effluent from the MWRP. The treatment scheme for the MWRP includes bar screens, primary clarification, activated sludge, secondary clarification, and coagulation with dual media filtration as the polishing step followed by chlorine disinfection. Sand Canyon also captures a significant amount of stormwater runoff, which contributes silt to the reservoir. The non-potable system is supplied with MWRP water during low demand periods. During these low demand periods, excess treated effluent can be conveyed to Sand Canyon Reservoir and the Zone A/Rattlesnake Reservoir complex. Water from both of these reservoirs is used to supplement reclaimed water when system demands increase during the summer. An additional source of supply to Rattlesnake Reservoir is untreated surface water from the Irvine Lake Pipeline and minor amounts of stormwater runoff.

IRWD has experienced several water quality problems characterized as turbidity, odor, and algae growth within the two reservoirs. At times, turbidity levels caused by algae growth and natural runoff have exceeded 100 nephelometric units (NTUs) with apparent color units (CUs) in excess of 1,000 CUs. Dissolved sulfide levels from the reservoirs have exceeded 20 mg/L, resulting in severe odor complaints. Coliform levels have also been found to exceed 23 most probable number (MPN) in the distribution system. These problems have resulted in 1) high color levels in the distribution system and, in turn, at high rise facilities, 2) continued need for agricultural users to filter water to prevent drip emitter clogging, and 3) odor complaints from homeowner associations (IRWD and Black & Veatch, 1994).

A comparison of average color levels for both reservoirs and the MWRP source water effluent from data collected from 1991 through 1992 is given in Figure 4-2. Comparing color data of plant effluent and the reservoirs demonstrates that severe increases in color occur within the reservoirs. Color levels in the MWRP effluent are consistently below 35 apparent CUs, while Rattlesnake and Sand Canyon Reservoirs have color levels in excess of 50 CUs and as high as 450 CUs (Harvey MUDD College Engineering Clinic, 1993).

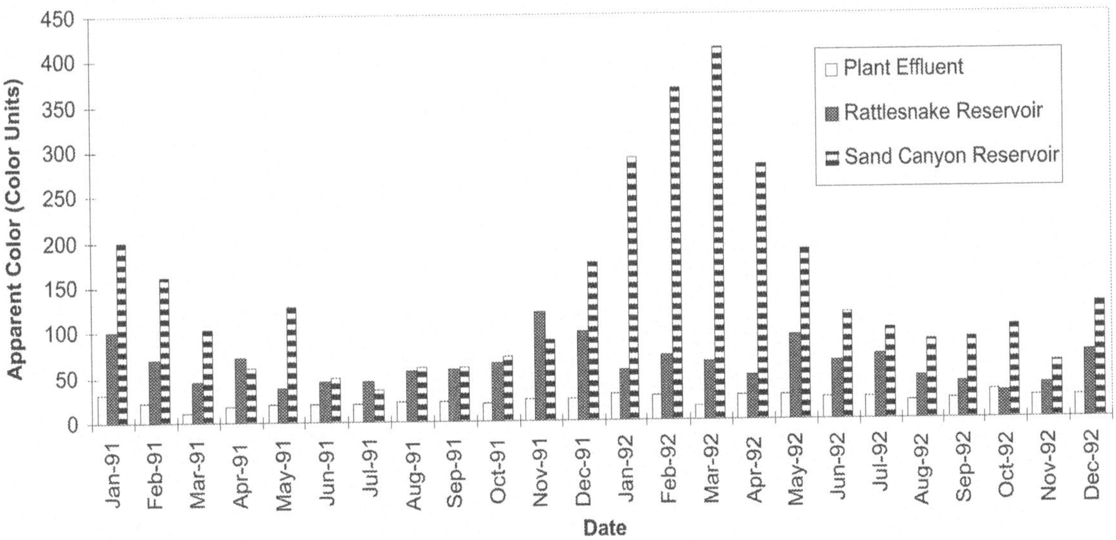

Figure 4-2. Comparison of Plant Effluent and Reservoir Color Levels

The Harvey MUDD study showed that color and turbidity levels followed a similar pattern. In particular, during wintertime storm events and through the following spring, color levels and turbidity levels were highest in Sand Canyon Reservoir. Average turbidity levels for the two year period were 28.4 and 5.7 NTU for Sand Canyon and Rattlesnake, respectively. A strong correlation was also found between color and turbidity (R=0.83) for the reclaimed water system. Water from Sand Canyon contained high levels of clay and silt, which was verified in a later study by IRWD personnel. The source of the silt problem could be attributed to the runoff and soil characteristics of the reservoir. It was confirmed in the 1994 IRWD and Black & Veatch study that most of the Sand Canyon watershed is comprised of fine clay sediments which upon entering the reservoir stay in suspension throughout the spring.

As shown on Figure 4-2, color levels in both reservoirs still exceed plant effluent levels during the spring, fall, and summer months, indicating that algal growth was a secondary cause of color and turbidity problems.

The physical characteristics of a reservoir and its surrounding watershed play a major role in determining the water quality in the reservoir. Current and future land use, rainfall, and runoff information is presented in the following discussion. Physical characteristics of the two reservoirs and in-reservoir water quality data collected from 1986 through 1991 are summarized in Table 4-4.

Both Sand Canyon and Rattlesnake Reservoirs are very small and would be defined as "shallow lakes" based on classical limnology definitions (mean depth < 33' (10 meters)). Shallow reservoirs also tend to eutrophy rapidly when the mean depth is less than 33' (10 meters) (Bernhardt and Clasen, 1985).

Temperature and DO profiles collected from 1986-87 for both reservoirs demonstrated that direct stratification occurs in early spring with destratification occurring from the fall through the early spring. During the summer, when the reservoirs stratify, anoxic conditions occur, resulting in H_2S formation and subsequent odor production.

Table 4-4. Watershed and Reservoir Characteristics

Parameter	Rattlesnake	Sand Canyon
Physical		
Storage Volume (acre-feet)	1,440	768
Surface Area (acres)	60	42
Mean Depth (feet)	24	18
Watershed Area (acres)	1,293	4,288
Soil Characteristics	No data	90%+ fine clay sediments
Temperature Stratification	Spring-summer	Spring-summer
Dissolved Oxygen	<5 mg/L May-September at 3 meter depth and below	<5 mg/L May-September at 3 meter depth and below
Nutrients		
Total Phosphorous (mg/L)	1.5 to 24.5	5.7 to 15.3
Total Nitrogen (mg/L)	2.5 to 6.3	3.1 to 11.4
Chlorophyll *a* (μg/L)	9.4 to 165	38 to 82
Turbidity (NTU)	1 to 11	1 to 35
Algae (organisms/mL)	3,300 to 77,400	5,500 to 542,000

As reported in Table 4-4, Sand Canyon Reservoir has a much larger watershed than Rattlesnake Reservoir. At the time this manuscript was written, the Sand Canyon watershed was used primarily for pasture, with a small portion of low-medium density residential. Rattlesnake Reservoir watershed area was used for agriculture purposes. Average rainfall is about 15" per year for both reservoirs, but, due to the size of the Sand Canyon Reservoir watershed and the character of the soils in it, rainfall plays a major role in water quality issues. The Sand Canyon watershed is comprised of fine clay sediments that are highly erodible, and during storm events this material contributes to turbidity and discoloration of the reservoir water.

The nutrient data in Table 4-4 are a range of average nutrient concentrations for surface samples from both reservoirs. Wide variability in concentration levels occurs, which is primarily a function of the source of nutrients (reclaimed water and runoff). Reclaimed water ortho-phosphate levels averaged 10.8 and 4.3 mg/L during 1991 and 1992. Total phosphorous levels from runoff to Sand Canyon Reservoir ranged from 0.3 to 0.9 mg/L. Higher levels in the reservoirs can probably be attributed to the release of nutrients at the water/sediment interface. The average phosphorous levels for both Sand Canyon Reservoir and Rattlesnake Reservoir surface samples ranged from 5.7-15.3 and 1.5-24.5 mg/L, respectively, which far exceed the requirements for eutrophic states in natural lakes of 0.1 mg/L. Nitrogen levels follow a similar pattern. Although both reservoirs have characteristics of eutrophication, normal theories of nutrient dynamics and morphology relationships cannot be expected to apply directly to them because of the significant differences between nutrient concentrations in reclaimed water and natural lakes.

A by-product of excessive nutrients in the water column is high algal or biomass production. Chlorophyll *a* was measured as an indicator of algal production along with algal measurements. In general, chlorophyll levels were higher in the summer for both reservoirs with higher levels in Sand Canyon Reservoir. Higher algal counts were recorded for Sand Canyon Reservoir (5,500-542,000 organisms/mL) than for Rattlesnake Reservoir (3,300-77,400 organisms/mL). The turbidity data also follow the same trend as the algal data, with higher turbidity levels in Sand Canyon Reservoir (1-35 NTU) than for Rattlesnake Reservoir (1-11 NTU).

In summary, degradation of water quality in the two reservoirs was documented. Water quality degradation, (i.e. higher turbidity, color, and algal levels) was more severe in Sand Canyon Reservoir due to its shallowness, its larger watershed, and the resulting contribution of fine clays during storm events. As a result of these water quality problems, IRWD implemented a watershed management plan.

4.8 Water Quality Problems and Reservoir Management

Complaints regarding odor, color, and turbidity are indicators of water quality problems in reclaimed water reservoirs. These include excessive algae growth, oxygen depletion, H_2S odor, and high turbidity. The conditions of excessive nutrients, algae, oxygen depletion, odors, and turbidity are interrelated.

4.8.1 Factors Limiting Growth

Reclaimed water reservoirs provide ideal conditions for algae growth: ample nutrients from the reclaimed water and sunlight. Runoff also supplies nutrients, although concentrations are usually an order of magnitude less than reclaimed water. The reclaimed water and runoff are considered to be external sources of nutrients. Internal sources include recycling of nutrients as algae sink and decay, and release of nutrients from the sediments under anoxic conditions.

Both phosphorus and nitrogen are necessary for algal growth. In most natural lakes, phosphorus is the limiting nutrient. As phosphorus is depleted, algae stop growing. In Rattlesnake and Sand Canyon Reservoirs, the minimum measured phosphorus concentration was 0.5 mg/L, which is at least 10 times higher than the amount needed to control algae growth in natural systems. In addition, orthophosphorus, the bioavailable form of phosphorus, comprises a larger fraction of the total phosphorus than would normally be found in reservoirs.

Nitrogen can also be the limiting nutrient, especially in eutrophic lakes. Inorganic nitrogen species consisting of ammonia and nitrate are the bioavailable forms. In general, if TIN falls below 0.1 mg/L, nitrogen may limit growth. If TIN is above about 0.4 mg/L, nitrogen would not limit growth. In Sand Canyon Reservoir, TIN concentrations were less than or equal to 0.4 mg/L on three occasions. In Rattlesnake Reservoir, these concentrations were reached four times. Although average values were considerably higher in both reservoirs, the TIN concentrations indicate that, at times, nitrogen concentrations may be approaching limiting levels. There was no consistent pattern showing when TIN concentrations would be low.

4.8.2 Oxygen Depletion

As a reclaimed water reservoir stratifies, the lower layers are effectively cut off from the surface and its supply of oxygen. As algae die, sink, and decay, they consume oxygen. The temperature and oxygen profiles for reclaimed water reservoirs indicate that the oxygen depletion rate in the lower layers is high during times of stratification. Oxygen depletion leads to the formation of H_2S, the major cause of odors.

In addition to generating odors, oxygen depletion also greatly increases the rate of release of phosphorus to the overlying water column. The released phosphorus is not immediately available to algae growing in the upper layers. However, at times of fall turnover, when the water column is mixed, phosphorus will be distributed throughout the water column. The additional supply of phosphorus may not be important related to the amount of phosphorus contained in the reclaimed water. However, it is an important consideration for any management alternative that would try to make phosphorus the limiting nutrient.

4.8.3 Turbidity

Turbidity is the measure of the clarity of the water. Several factors affect turbidity, including suspended sediments, colloidal material, algae, and color. The turbidity is measured by determining light transmission using standard light sources. Reclaimed water treatment goals typically have a turbidity goal which are based on the level of treatment. Turbidity levels of reclaimed water plant effluent are typically very low, less than 2 NTUs, as shown in Table 4-1. Reservoirs that store reclaimed water which have natural bottoms would be expected to have higher turbidity levels due to algae growth, re-suspension of sediments into the water column, and runoff. Turbidity control measures need to be considered to minimize post treatment of reclaimed water in storage reservoirs.

CHAPTER 5.0

RESERVOIR MANAGEMENT STRATEGIES

5.1 Overview

Reservoir management includes a set of strategies designed to improve the quality of water in the reservoir. A number of management strategies currently used in lakes and freshwater reservoirs to enhance and improve water quality may be applicable to reclaimed water storage reservoirs as well. Substantial information on management strategies was obtained from the literature survey. However, it is extremely important to note that much of the available information on reservoir management and restoration comes from the study of natural lakes. Reservoirs act differently than lakes, and reclaimed water reservoirs are significantly different than natural lakes. These differences must be considered carefully when reviewing results of programs conducted elsewhere.

This chapter starts by providing guidance on how to evaluate and select management strategies that could be effective in improving water quality in reservoirs. Next, information is presented on several strategies that are the most likely to be effective in reclaimed water storage reservoirs. Information presented for each includes a description of the technique, the problems addressed, effectiveness in reclaimed water reservoirs, and other considerations. This information is summarized in Table 5-1 at the end of the chapter.

5.2 Evaluation and Selection of Management Strategies

The effectiveness of any management strategy will depend to a large extent on site-specific conditions, including the physical characteristics of the reservoir, runoff from the watershed, the initial quality of the reclaimed water, and reservoir operations. Therefore, a management strategy may be very useful in one reservoir and ineffective in another. This section describes an approach that can be used by utility managers to evaluate and select management strategies that could be effective in improving water quality in their reservoirs. It includes identifying problems, defining the physical and operational characteristics of the reservoir, screening potential strategies, evaluating alternatives, and selecting feasible management strategies. The approach presented is essentially the same as that used to conduct diagnostic/feasibility studies as part of the Clean Lakes Program (Olem and Flock, 1990.)

5.2.1 Identify Problems

In previous chapters, the criteria and guidelines used to regulate the use of reclaimed water, the water quality changes that occur in open storage reservoirs, and the problems encountered by reclaimed water utilities have been presented and discussed. In general, the water quality problems encountered in reclaimed water storage reservoirs are a direct or indirect result of excess nutrients and algae, low dissolved oxygen (DO), high turbidity, odors, and color. The first step in selecting management strategies is to identify the specific problems encountered in the reservoir and the sources of the problems. Monitoring data and customer complaints can provide useful information in this process. A reservoir water quality model can also be useful to identify sources of water quality problems. Information on defining water quality problems and on developing and conducting water quality monitoring programs is available from several sources (e.g., Olem and Flock, 1990; Wedepohl et al., 1990; Cooke et al., 1993; and Cooke and Carlson, 1989). Water quality models are discussed in Chapter 6.0 of this report.

5.2.2 Define Reservoir Characteristics

Defining reservoir characteristics is an important element in both identifying the sources of water quality problems and in the evaluation of alternatives. Such characteristics include:

- Physical measurements of the reservoir, such as surface area, volume, mean depth, and thermocline depth.
- Water balance, including inflows of reclaimed water, inflows from the watershed, and hydraulic residence time.
- Nutrient balance, including the loading rates from the reclaimed water, the watershed, and internal sources.
- Reservoir operations, including the timing of inflow and withdrawals, drawdown cycles, and any existing water quality improvement techniques.

5.2.3 Screen Potential Strategies

A comprehensive list of management alternatives that could mitigate water quality problems in reservoirs has been identified. These strategies include destratification/artificial circulation, hypolimnetic aeration/oxygenation, hypolimnetic withdrawal, dilution, alum treatment, dredging, wetlands treatment, biomanipulation, targeted plant growth, advanced treatment for nutrient removal, copper sulfate, and watershed management. These strategies have all been implemented in freshwater lakes and reservoirs around the country with varying degrees of success. The next step in evaluating and selecting management strategies is to screen all potential strategies to determine the potentially feasible alternatives; others should be dropped from further consideration. The screening evaluation should consider the specific problems and sources that the alternative addresses, the reservoir and watershed characteristics, and reservoir operations.

5.2.4 Evaluate Alternatives

The final step in selecting management strategies is to evaluate each strategy. The evaluation should consider:

- Effectiveness of the strategy in improving water quality.
- Implementation issues including permitting and public acceptance.
- Operation and maintenance issues.
- Costs.

5.2.5 Strategies Dropped from Consideration

Some of the strategies named above, including hypolimnetic withdrawal, dilution, biomanipulation, and targeted plant growth are unlikely to be effective in improving water quality in reclaimed water storage reservoirs. These strategies are briefly described here, but are not considered further in this chapter.

Hypolimnetic withdrawal is used to remove nutrient rich bottom water from reservoirs. Bottom water is removed, and the reservoir is filled with lower concentration runoff. This alternatuve is useful in reservoirs with multilevel withdrawal ports. It would not be effective in most reclaimed water reservoirs because the entire water column is usually nutrient-rich, especially if the majority of the water entering the reservoirs is reclaimed water. Hypolimnetic withdrawal would be a possible strategy only in the unusual circumstance where a reservoir is stratified and the nutrient-rich reclaimed water resides near the thermocline and near the bottom. For these reasons, this alternative was dropped from further consideration.

Dilution has been used as an effective measure to improve water quality in lakes (Cooke and Carlson, 1989; Cooke et al., 1993). Dilution can be effective if a supply of low-nutrient water can be introduced into the reservoir to dilute higher nutrient concentrations and provide flushing. Dilution is usually effective only where a large supply of low-nutrient water is available, for example, from groundwater or a river. In most areas with reclaimed water storage reservoirs, water is scarce, and supplies for dilution are not available. Therefore, dilution was not considered to be a feasible management strategy for reclaimed water storage reservoirs.

The objective of the biomanipulation strategy is to control algal growth by manipulating the aquatic food chain to increase the population of zooplankton (Cooke et al., 1993). Zooplankton are small planktonic animals that filter large volumes of water and graze on algae, bacteria, and bits of organic matter. A high biomass of large-bodied zooplankton can effectively control algal growth. However, not all species of algae are palatable to zooplankton, including certain colonial and filamentous blue-green algae associated with eutrophic reservoirs.

In many cases, biomanipulation has been successful. By manipulating the food chain and replacing planktivorous fish, such as blue gill, perch, shad, and alewife, with predatory, piscivourous fish, such as walleye or large-mouth bass, the population of zooplankton should increase (Brooks and Dodson, 1965). However, long-term maintenance of the fish population can be difficult. To effectively manage reservoir fisheries, the entire ecosystem at all trophic levels must be understood and considered in management decisions (Noble, 1986).

Biomanipulation seems to work best in lakes with a large percentage of shallow areas, a well established macrophyte community, a total phosphorus concentration of less than 200 micrograms per liter (μg/L), and green algal and diatoms as the predominant species of phytoplankton. The lake characteristics for most reclaimed water reservoirs do not match those required for successful implementation of biomanipulation. The phosphorus levels are well above the suggested limits, and blue-green algae often predominate. Also, if operating plans call for significant drawdown of the reservoirs, it may not be possible to support a macrophyte community or provide sufficient habitat for the fish population. Therefore, it appears that this technique would not be feasible technically for reclaimed water storage reservoirs.

Targeted plant growth is a management strategy whereby selected macrophytes are introduced to the reservoir to remove nutrients from the water and to shade algae to prevent their growth. Although moderately effective at controlling algal growth in some situations, several problems are associated with implementing this alternative in reclaimed water reservoirs. It could be difficult to establish a good macrophyte community because of the normal operating procedure of drawdown that brings many reservoirs to very low levels each year. Drawdown would likely kill a large portion of the macrophyte biomass. In fact, water level drawdown is a recommended method for controlling nuisance macrophytes. If the macrophytes could be established, the plant growth would require periodic harvesting and disposal of plant material. Also, it is possible that introduced plants could grow at an undesirable rate and become a nuisance. This alternative was considered infeasible and was dropped from further consideration.

5.3 Destratification/Artificial Circulation

5.3.1 Description of the Technique

As discussed in Chapter 4.0, many reservoirs are deep enough to thermally stratify. After stratification is established in the reservoirs, the oxygen supply to the hypolimnion is effectively cut off. The bottom waters become anoxic, and hydrogen sulfide forms, causing odors in the reclaimed water. In addition, phosphorus is released from the bottom sediments under anoxic or low oxygen conditions, adding to the nutrient loading of the reservoirs.

The reservoirs could be destratified by pumping compressed air through a series of perforated pipes into the bottom water of the reservoirs. As the air bubbles rise, they would establish vertical mixing in the reservoirs. The compressed air itself would not add much oxygen directly to the water. However, the artificial recirculation would allow for an increased exchange of oxygen at the surface. Compressed air systems only need to cover a portion of the lake bottom. Once the artificial circulation was established, it could affect the entire reservoir.

5.3.2 Problems Addressed

Destratification would prevent anoxic conditions from occurring in the reservoirs, would address the odor problems caused by formation of hydrogen sulfide, and should also control the release of phosphorus from bottom sediments. Destratification may affect the growth of algae by causing them to spend part of their time in darkness (Fast, 1979). Very buoyant species would not be affected by this method because their buoyancy would negate the effects of downward mixing.

5.3.3 Effectiveness for Reclaimed Water Storage Reservoirs

Numerous case studies of aeration systems are available. In addition, several reviews have been conducted to evaluate this restoration alternative (e.g. Pastorak, Lorenzen, and Ginn, 1982;

Cooke and Carlson, 1989). These reviews show that most systems are able to improve DO concentrations. Proper sizing of the system is essential; most failures are due to input of insufficient energy to maintain isothermal conditions. Cooke et al. (1993) presented an updated review and evaluation of aeration systems and cited at least 50 cases where aeration systems have been installed. Dissolved oxygen concentrations improved in 80% of the cases.

Many systems are installed to control nutrient inputs from bottom sediments. Their success rate is mixed. In a survey of mixing applications, nutrient concentrations in about one-third of the lakes increased, while one-third experienced a decrease in nutrient concentration; the remaining one-third showed no change or a questionable response (Cooke and Carlson, 1989). Nutrient and algae increases are more likely in lakes that are nutrient-limited, because mixing makes nutrients from the hypolimnion available for growth (Cooke et al., 1993).

In some cases, algae populations have decreased after destratification, while in other cases they have increased or remained unchanged. Algae concentrations can decrease because of lower nutrient concentrations or because of the increased depth of the mixed layer and the greater amount of time they spend in darkness. Destratification is not as effective in reducing the concentrations of blue-green algae because of their buoyancy.

An aeration system installed at Sand Canyon Reservoir, a reclaimed water storage reservoir operated by the Irvine Ranch Water District (IRWD), has been very effective at alleviating the low DO and odor problems. It is likely that aeration systems installed at other reclaimed water storage reservoirs would also be effective in controlling the problems of low DO and odors. The effects of aeration on controlling internal nutrient loading, turbidity, color, and algae are uncertain. It is possible that oxygenating the bottom waters of the reservoirs may significantly reduce internal nutrient loading. However, if external sources of nutrient loading are not controlled, a reduction in the internal sources would likely not be noticed.

5.3.4 Other Considerations

A common negative impact of destratification is noise complaints because of the air compressors. The significance of this problem will be related to the proximity of residences to the reservoirs. The problem could be mitigated by using sound-proof housings for the compressors.

5.4 Hypolimnetic Aeration/Oxygenation

5.4.1 Description of the Technique

Destratification is an effective technique for alleviating low DO problems in reservoirs. However, in some cases, it is desirable to maintain the thermocline to preserve cold water fishery resources or prevent hypolimnetic nutrients from mixing into the epilimnion. In addition, it is not practical to destratify some very large reservoirs. In these cases, hypolimnetic aeration can be used to elevate DO concentrations in the hypolimnion. Hypolimnetic aeration adds oxygen to the bottom layers of a reservoir without establishing artificial vertical circulation, thereby maintaining the thermocline. Air and commercial oxygen have both been used in hypolimnetic systems. In many cases, the use of commercial oxygenation is a better choice than air (Speece, Nirmalakhandan, and Tchobanoglous, 1990).

5.4.2 Problems Addressed

Hypolimnetic aeration or oxygenation would address the problems of low DO in reclaimed water storage reservoirs. In addition, this aeration technique could reduce phosphorus release from sediments under anoxic conditions.

5.4.3 Effectiveness for Reclaimed Water Storage Reservoirs

There are several designs for hypolimnetic aeration systems. One technique used in Tennessee Valley Authority (TVA) reservoirs is to discharge fine oxygen bubbles at the bottom of the reservoir. If there is sufficient depth (about 30 m), the oxygen will be dissolved completely into the water prior to reaching the thermocline. In this way, the stratification is maintained in the reservoir. Until recently, full or partial air-lift systems were the most popular design for hypolimnetic aeration. With an air-lift system, water is mixed with air, the water and air bubbles are separated, the air is released at the surface, and the water is discharged at depth (Cooke et al., 1993). A newer method is to use a Speece Cone and establish horizontal flow of highly oxygenated water (Speece, 1994 and 2000). The Speece Cone is capable of maintaining a thermocline in shallower lakes. The Speece Cone has been used successfully in several locations and should be effective for use in reclaimed water reservoirs (Cuff, 1993; Speece, 1994).

In most reclaimed water storage reservoirs, it will not be necessary to maintain the thermocline. However, using a fine bubble oxygen system or the Speece Cone would be an effective way to increase DO concentrations in the hypolimnion of the reservoir.

5.4.4 Other Considerations

Using commercial oxygen instead of compressed air would alleviate noise concerns. Oxygenation would avoid the noise problems. In addition, it has been shown that oxygenation tends to strip dissolved nitrogen from the water, allowing it to be released as nitrogen gas.

5.5 Wetlands Treatment

5.5.1 Description of the Technique

Constructed wetlands are shallow pools that create conditions for the growth of emergent vegetation. Wetlands are constructed in the watershed near the reservoir to treat reclaimed water. In a conventional wetlands system, both nitrogen and phosphorus are removed by the aquatic vegetation, which must be harvested periodically to achieve a long term reduction in nutrients. If unharvested, the plants eventually decay and release nutrients back to the water, resulting in little long-term improvement in water quality.

Another type of wetland system is designed to reduce nitrogen concentrations to levels where nitrogen would be the limiting nutrient. The wetland would utilize denitrification to convert inorganic nitrogen to nitrogen gas, which is removed from the system to the atmosphere. A relatively small amount of phosphorus is removed by the wetlands vegetation. A major disadvantage of this method is that it relies on removing nitrogen to a level low enough so that nitrogen rather than phosphorus controls the growth of algae. Because some species of blue-green algae can obtain nitrogen from the atmosphere when nitrogen in the water is depleted, the nitrogen-limiting wetland may not always be effective in controlling algae. Given this limitation, conventional wetland systems are often preferred.

5.5.2 Problems Addressed

Wetlands systems have been used effectively to treat stormwater runoff and wastewater. The wetlands would control algae growth by removing both nitrogen and phosphorus from reclaimed water and the lake water. Wetlands are more effective in removing nitrogen than phosphorus (U.S. EPA, 1988).

5.5.3 Effectiveness for Reclaimed Water Storage Reservoirs

Wetlands are increasingly being used to treat wastewater and stormwater runoff. However, in many cases, monitoring programs are inadequate to determine the efficiency of the systems and the mechanisms responsible for removal (Landers and Knuth, 1991). The mechanisms by which phosphorus and nitrogen are removed from wastewater by wetlands and the relationship between nutrient application rates and removal by wetlands were described by Nichols (1983); nine case studies were summarized. Total phosphorus removal ranged from 20% to 96%, while total nitrogen removal ranged from 1% to 96%. The systems with low removal rates had been in operation for over 50 years, indicating that wetlands soils may become saturated with nutrients.

Although wetlands can be a highly effective means to reduce nutrient inputs to the reservoirs, this alternative would be land intensive. For phosphorus removal, about 100 acres of wetlands would be required to treat 1 mgd of reclaimed water with a phosphorus concentration of 3.5 milligrams per liter (mg/L). Nitrogen removed by the wetlands is dependent on the residence time in the system, and land requirements for nitrogen removal would also be significant. However, if nitrification/denitrification were first implemented, the use of wetlands could be a feasible alternative; the requirements for land would be reduced because less treatment for nitrogen removal would be required in the wetlands (see discussion in Section 5.8).

Overall, it appears that wetlands treatment for reclaimed water would not be feasible in most cases because of the land requirements. However, if nutrient removal is first implemented, wetlands could be used to further reduce concentrations in the reservoirs and would aid in controlling algae concentrations.

5.5.4 Other Considerations

Some wetlands require a constant supply of water, so provisions should be made to pump reservoir water through the wetlands during times when reclaimed is not supplied or available. Recycling reservoir water through the wetlands would provide additional nutrient removal.

If wetlands were used to treat reclaimed water inputs to the reservoir, additional measures to treat internal sources or other external sources of nutrients may be necessary to achieve significant improvements in water quality.

Periodic mowing or harvesting of vegetation is required, as well as periodic removal and replacement of nutrient saturated soils.

5.6 Copper Treatment

5.6.1 Description of the Technique

Algicides are applied to reservoirs to control nuisance algal blooms that result from high in-lake nutrient concentrations. Copper can be a very effective algicide, but its effectiveness is temporary, and it can cause negative impacts on other components of the ecosystem. The cupric ion is the toxic form of copper, and water chemistry plays an important role in determining the availability of this form. McKnight, Chisholm, and Harleman (1983) discussed the various forms of copper and the processes that affect the concentration of cupric ion in water. These processes include inorganic complexation, precipitation, complexation with humic and fulvic compounds, adsorption onto clay particles, and biological uptake. Alkalinity and pH also have a significant effect on the availability of the cupric ion.

Because water chemistry is different in every reservoir, it is difficult to establish guidelines for the proper dosage of copper sulfate applications. Copper is less toxic in hard water; therefore, higher dosages of copper sulfate are required to be effective. Copper sulfate is perhaps the most common form of copper applied to reservoirs. However, complexing or chelating agents are often added to keep the copper in solution and increase its effectiveness. For example, cutrine, a copper and organic nitrogen derivative, contains a chelating agent that allows it to remain in solution longer than copper sulfate.

5.6.2 Problems Addressed

Copper sulfate is used as a temporary measure to control algae. It does not address the causes of algal blooms.

5.6.3 Effectiveness for Reclaimed Water Storage Reservoirs

Case studies for the application are provided in McKnight, Chisholm, and Harleman (1983) and Casitas Municipal Water Department (1987). However, a more applicable example is the testing conducted by the IRWD on two reclaimed water storage reservoirs: Rattlesnake and Sand Canyon Reservoirs (IRWD and Black & Veatch, 1994). Copper sulfate was applied at a concentration of about 1 mg/L as recommended by the manufacturer. The number of pounds required to obtain the recommended dosage was calculated based on the storage volume of the reservoir. The treatment was applied by loading copper sulfate into burlap sacks, hanging the sacks over the side of a boat into the water, and cruising slowly around the reservoir. Personnel applying the treatment noted that the powder form of copper sulfate dissolved slowly allowing for a fairly uniform mixture within the reservoir. Sampling was conducted in 1990, 1991, and 1992. In 1992, algal samples were collected one hour before and one hour after sampling and also five days after sampling. The pattern of different species counts varied considerably. After five days, some were at or above their pretreatment levels. However, many of the species were well below pretreatment concentrations. Overall, there was a 20% reduction in algae counts after five days. Unfortunately, the water chemistry of Rattlesnake and Sand Canyon Reservoirs has high alkalinity and high pH, which are not conducive to maintaining copper in solution. Previous applications of copper sulfate in these reservoirs had been somewhat effective.

The use of copper sulfate has been a somewhat effective strategy to control algal growth in reservoirs for short periods of time. For copper sulfate to be most effective, it should be applied during the log growth phase. Monitoring algal counts is critical to the successful application of copper sulfate. It is unlikely that copper sulfate can be used as the only control measure for algal growth in the reservoirs to attain the water quality desired by most utility managers. However, copper treatment would likely be part of an overall water quality management plan to temporarily control growth of algae.

5.6.4 Other Considerations

Laboratory tests have shown that copper sulfate is toxic to zooplankton. Although there are questions about the applicability of these tests to real systems, concerns about copper toxicity remain. In addition, copper sulfate treatment could elevate the concentration of copper in sediments leading to disposal problems if the reservoirs are dredged. Because of environmental concerns, copper sulfate applications are not allowed in some areas. State and local regulations should be checked prior to use.

5.7 Advanced Waste Treatment for Phosphorus Removal

5.7.1 Description of the Technique

Adding advanced waste treatment for phosphorus removal prior to discharging the reclaimed water into the reservoir is an important alternative for the control of algal growths. Phosphorus removal can be performed using biological nutrient removal, chemical precipitation, effluent filtration, and combinations of these processes. Effluent concentrations of phosphorus in the range of 0.15 to 0.25 mg/L are achieved at several plants in Canada (Barnard, 2002). The Durham wastewater treatment plant in Oregon achieves its permit limit of 0.07 mg/L total phosphorus using a combination of enhanced biological phosphorus removal, chemical precipitation and effluent filtration (Benisch et al., 2001).

5.7.2 Problems Addressed

Phosphorus removal would address the problem of excessive phosphorus concentrations in reclaimed water storage reservoirs. In phosphorus limited reservoirs, its removal should also control algal growth.

5.7.3 Effectiveness for Reclaimed Water Storage Reservoirs

Several reviews have been published on the response of lakes to reductions in external phosphorus loading. Cooke et al. (1993) summarized these and discussed several case studies that reflect the range of results observed. In general, the case studies showed that lakes do respond, but often the response is slower or less than expected. None of the case studies involved reclaimed water storage reservoirs, but some of the observations are applicable to this project.

Responses can be classified into three types:

- Type 1—Sufficient to change trophic state.
- Type 2—Reduction in total phosphorus and chlorophyll *a*, but insufficient to change trophic state.
- Type 3—Small or no obvious improvement in total phosphorus or chlorophyll *a*.

Cullen and Forsberg (1988) reviewed the response of 43 lakes that experienced reductions in external phosphorus loading. Fifteen (35% of the lakes) were classified as Type 1, 9 (21%) were classified as Type 2, and 19 (44%) were classified as Type 3. Based on the results of the studies, it was found that the percent reduction in measured lake concentration as phosphorus was less than the percent reduction in inflow concentration. The reduction was usually due to internal nutrient loading. Lakes classified as Type 2 or Type 3 had average lake concentrations in excess of 100 µg/L. However, it was also found that, although total algal biomass might not change significantly, shifts in species composition can occur as a response to phosphorus reduction.

Lake Norrviken, Sweden, one of the case studies discussed in Cooke et al. (1993), is an example of a hypereutrophic lake that fits the Type 2 classification defined above. After external phosphorus reduction, the lake experienced internal reductions of phosphorus and chlorophyll *a* of 67% and 43%, respectively. In addition, dominant species shifts occurred so that algal blooms were less of a nuisance. The project was considered a success even though the lake remained hypereutrophic.

Phosphorus removal could reduce the incoming phosphorus concentration to about 0.08 mg/L. The objective of this strategy is to force phosphorus to become the limiting nutrient. Internal sources would be unaffected by this technique. Phosphorus release from bottom sediments under reducing conditions could still be sufficient to cause algae problems. However, other strategies including dredging, alum treatment, and aeration could be implemented to control internal sources.

5.7.4 Other Considerations

In many cases, only a portion of the reclaimed water is stored in reservoirs. However, all of the water would be treated, because in most cases there is no way to separate the water entering the reservoirs from that entering the rest of the distribution system. Cost for phosphorus will be a major consideration in determining if it is feasible for a particular facility.

5.8 Advanced Waste Treatment for Nitrogen Removal

5.8.1 Description of the Technique

Nitrogen removal prior to the reclaimed water entering the reservoirs is another alternative for the control of aquatic growth. The goal would be to reduce total inorganic nitrogen to levels where nitrogen would be the limiting nutrient. Nitrogen removal can be achieved in several different ways.

5.8.2 Problems Addressed

The objective of this strategy is to reduce nitrogen concentrations and attempt to force nitrogen to be the limiting nutrient. Algae concentrations would be controlled through nutrient limitation.

5.8.3 Effectiveness for Reclaimed Water Storage Reservoirs

No case studies were found that are applicable to this alternative. While nitrification/denitrification has been installed at many treatment plants to improve water quality, the concern is usually ammonia toxicity and not eutrophication. To be nitrogen limited, total inorganic nitrogen (TIN) concentrations in the reservoir should be less than 0.1 mg/L. Even with a 50%

reduction in ammonia concentrations, concentrations of TIN in reclaimed water could still be substantial. The presence of nitrogen is complicated by the ability of some algae to fix nitrogen from the atmosphere, thereby converting nitrogen gas to inorganic forms.

By itself, the removal of nitrogen may not be a feasible management strategy. However, combined with wetlands treatment, it could be very effective and feasible. The wetlands system would be used for denitrification to convert inorganic nitrogen to nitrogen gas, which is removed from the system to the atmosphere. The amount of nitrogen removed by the wetlands is dependent on the residence time in the system. For example, reduction of TIN from 8 mg/L to 0.1 mg/L would require about seven acres of wetlands per 1 million gallon per day (mgd, 0.044 m^3/s) of reclaimed water treated (IRWD and Black & Veatch, 1994).

5.8.4 Other Considerations

In many cases, only a portion of the reclaimed water is stored in reservoirs. If only a portion of the wastewater needs to be treated, a sidestream of effluent would need to be diverted for treatment. Cost for this alternative will be a major consideration in determining if it is feasible for a particular facility.

5.9 Dredging

5.9.1 Description of the Technique

The primary objective for sediment removal in reservoirs is controlling internal nutrient sources. Dredging can remove the top layer of nutrient-rich sediments accumulated in the reservoir. Dredging is also used in some cases to remove nuisance macrophyte growth. Increasing the available storage in the reservoir would also be a benefit of sediment removal. In most reservoirs, sediment removal is accomplished using a mechanical or hydraulic dredge. However, in some cases, reservoirs can be drained, allowed to dewater, and the sediment removed by bulldozer.

5.9.2 Problems Addressed

Dredging would remove the surface layer of nutrient-rich sediments, thereby reducing the input of phosphorus under anoxic conditions. Nuisance macrophyte growth can also be removed with dredging. An added benefit would be an increase in storage volume.

5.9.3 Effectiveness for Reclaimed Water Storage Reservoirs

General reviews of dredging programs and projects are provided by Dunst (1981), Peterson (1982), and Cooke et al. (1993). There are many case studies of dredging projects, but few document hypereutrophic lakes where sediment removal for control of internal cycling was performed. Two specific case studies are discussed below.

5.9.3.1 Lake Trummen, Sweden

Lake Trummen, Sweden, is a well-documented case of sediment removal to control internal loading of nutrients and to control macrophyte growth (Cooke et al., 1993). Lake Trummen has an area of 400 hectares (ha) and, prior to restoration, had an average depth of 1.1 m and a maximum depth of 2 m. The lake had received various amounts of domestic and industrial wastes over a long period of time. Even after wastewater diversion, the lake experienced serious water quality problems because of internal nutrient loading. Sediment concentrations were mapped to provide a vertical and horizontal distribution of phosphate and ammonium ion nutrient concentrations. It was determined that approximately 40 centimeters (cm) of fine nutrient-rich sediments had accumulated over a period of 25 years. Nutrient release rates were also measured for surface and underlying sediments. The release rates for the surface sediments were greater than for the underlying sediments for both aerobic and anoxic conditions.

Sediment removal reduced the phosphate concentration in the surface sediment from about 2.4 mg/L to 0.1 mg/L. Total phosphorus concentration in the surface water of the lake decreased from 0.6 mg/L to less than 0.1 mg/L. At a phosphorus concentration of 0.1 mg/L, the reservoir would still be considered eutrophic; however, some water quality benefits were realized. The blue-green algal biomass decreased sharply, and some nuisance species disappeared. In addition, the benthic community recovered rapidly.

5.9.3.2 Lilly Lake, Wisconsin

Lilly Lake has a 37 ha surface area and maximum and mean depths of 1.8 m and 1.1 m, respectively. A detailed report on this dredging project is provided by Dunst et al. (1984). The lake had been filling in at a rate of 0.5 cm/year and, in one area, had accumulated 10.7 m of sediment containing masses of decaying plant material. The objective of the project was to deepen the lake to 6m by removing about 680,000 cubic meters (m^3) of sediment. The lake was dredged during the open water periods of 1978 and 1979.

The study monitored short-term and long-term impacts of the dredging project. Some short-term adverse impacts were observed, including an increase in TIN during the active dredging period. Total phosphorus also increased during dredging. Phytoplankton growth responded to the increased nutrient supply with an increase in the measured rate of primary productivity.

Although the main purpose of the dredging was to address macrophyte problems, long-term improvements in lake water quality were also observed. Total phosphorus concentrations after dredging were measured in the range of 0.014 to 0.18 mg/L, as opposed to the pre-dredging concentration of 30 to 60 μg/L. Nitrogen decreased, although slightly, after dredging. Dredging also eliminated the previous problems of low DO conditions and winter fish kills. Cooke et al. (1993) reported that, 10 years after dredging, water quality continued to remain good with no measurable increases in nutrient concentrations.

5.9.4 Other Considerations

Dredging could remove a major source of nutrients and sediment oxygen demand in reclaimed water storage reservoirs. However, without reducing the nutrient concentrations in the reclaimed water, dredging would not produce measurable results. The benefits of reducing internal sources of nutrients will be negated if the reservoirs continue to be filled with nutrient-rich reclaimed water.

Sediment testing should be conducted to define adequately the feasibility of this alternative for a specific reservoir. Tests should be conducted to determine sediment nutrient concentrations, release rates and the depth of nutrient-rich sediments. Sampling and analytical methods to characterize bottom sediments are discussed by Gambrell, Reddy, and Khalid (1983); Dunst et al. (1984); and Cooke et al. (1993).

Dredging would require a permit from the U.S. Army Corps of Engineers and possibly other state or local permits. Disposal of the dredged material may be difficult, especially if the sediments have elevated concentrations of metals or other pollutants. Dredging is also a relatively expensive strategy although it can have long-term effectiveness.

5.10 Alum Treatment

5.10.1 Description of the Technique

Chemical inactivation of phosphorus is a restorative measure used in lakes and reservoirs to prevent the internal cycling of phosphorus from sediments or to precipitate phosphorus from the water column. Internal phosphorus cycling may contribute a significant load of phosphorus to the water column to support algal blooms. Iron or aluminum salts, such as ferric or aluminum sulfate (alum), are applied directly to precipitate phosphorus from the water column by flocculation and to bind the phosphorus, which is adsorbed to reservoir sediments, to the aluminum hydroxide. The floc formed while precipitating phosphorus from the water column settles over the sediments and creates a chemical barrier, which prevents the release of phosphorus from the sediments.

Alum is normally applied as a liquid to the surface of the reservoir. Because it is heavier than water, the alum sinks to the bottom, carrying with it the soluble phosphorus and even particulate matter, such as algae. Stripping phosphorus from the water column results in an immediate improvement in water quality shortly after the alum is applied. The long term benefit comes after the alum settles to the bottom and forms a thin layer, which inactivates the phosphorus in the sediments, keeping it from returning to the water. Long term effectiveness of an alum treatment is dependent on keeping the alum blanket intact and on controlling external sources of phosphorus to the reservoir.

5.10.2 Problems Addressed

An alum treatment would reduce the internal recycling of phosphorus from bottom sediments and would precipitate phosphorus from the water column.

5.10.3 Effectiveness for Reclaimed Water Storage Reservoirs

Alum treatment could be used to reduce the input of nutrients from the reservoir sediments, which can be a significant source of phosphorus. In addition, reductions in nitrogen and metals to the reservoir could occur. To determine if an alum treatment would be effective for reclaimed water reservoirs, two important issues must be considered: (1) the concentration of nutrients in the reclaimed water, and (2) reservoir operations. If the nutrient concentrations of the reclaimed water are not reduced, and the reservoir is continuously filled with high-nutrient water, the benefits of an alum treatment would be negated. In this situation, the only time that alum could be an effective treatment would be if reclaimed water were added to the reservoir infrequently and alum treatment were conducted routinely. The feasibility of alum precipitation depends on the timing and cost of applications.

If nutrients are removed from the reclaimed water, alum treatment could be an effective way to control phosphorus release from reservoir sediments. However, the long term effectiveness of the treatment is dependent on keeping the alum blanket intact. If the alum blanket is disturbed, for example, by resuspension of sediment by wave action, its effectiveness will be significantly reduced. Some reservoirs undergo significant drawdown periodically. Alum treatment would be incompatible with this operation and would not be effective in these situations.

5.10.4 Other Considerations

A number of studies have shown that alum treatment can be effective in significantly reducing concentrations of phosphorus and chlorophyll *a* and in reducing turbidity. Although few long-term studies have been conducted, it is generally thought that alum treatment would be effective for three to five years. Monitoring for Kezar Lake, NH, showed that water quality improvements were still observed 10 years after an alum treatment of that lake (Tetra Tech, 1995). The long term effectiveness of an alum treatment is dependent on keeping the alum blanket intact and on controlling external sources of phosphorus to the reservoir. Reservoir operations must be considered carefully in the evaluation of the management strategy.

5.11 Watershed Management

5.11.1 Description of the Technique

Development of a watershed management plan can be used as a tool to reduce the loadings of nutrients and other pollutants to reclaimed water storage reservoirs. Watershed management involves the identification of all of the sources of pollutants and measures to control them. Measures to control pollutants can include source control programs or structural Best Management Practices (BMPs) to treat runoff from the watershed before it enters the reservoir. Development of watershed management plans and descriptions of structural and nonstructural BMPs are described in many sources (e.g., Robbins et al., 1991; Schueler, 1987). As examples, two types of BMPs that could be used to treat watershed runoff are discussed in this section: detention basins and in-stream treatment.

Wet detention basins have been used widely to retain runoff and trap sediments and associated pollutants. This management strategy would include constructing one or more wet detention basins in the watershed, preferably near the reservoir. Another option would be to convert an arm of the reservoir into a detention basin by constructing a small dam across the arm. The arm could be deepened to a depth of about 1m to provide additional storage volume and retention time.

In-stream treatment would consist of treating the watershed runoff as it enters the reservoir. Treatment could include the addition of alum or ferric chloride to flocculate sediment particles. The treated runoff water and residuals would be discharged into the reservoirs. A major drawback of in-stream treatment would be the build-up of residuals in the reservoir. An option to help control the build-up would be to convert an arm of the reservoir to a detention basin where the treated water would be discharged. The detention basin would confine the floc to a relatively small area while allowing it to settle. Ongoing maintenance of the basin would be required, but would be easier than dredging larger areas of the reservoir.

5.11.2 Problems Addressed

Implementing a watershed management plan could reduce external loading of nutrients and sediments from the watershed.

5.11.3 Effectiveness for Reclaimed Water Storage Reservoirs

The watershed areas of reclaimed water storage reservoirs vary widely. Some reservoirs have almost no contributing watershed area, while others have areas large enough to contribute a significant portion of the total inflow to the reservoir. If runoff from the watershed comprises a significant portion of the water balance of the reservoir, watershed controls may be effective in improving water quality in the reservoir. However, they will be ineffective if nutrient concentrations in the reclaimed water and from internal sources are not controlled.

Although the procedures used to develop watershed management plans may be similar, in many cases, the exact elements of the plan will depend on the site-specific nature of the watershed. A variety of watershed management techniques are available to reduce pollutant loadings from the watershed to the reservoir. Watershed BMPs have been shown to be effective in reducing loadings of nutrients to reservoirs and improving their water quality.

Detention basins are one way to treat runoff from the watershed to remove sediments and associated nutrients. The effectiveness of a detention basin is dependent on the size of the basin relative to the volume of runoff, and the removal rates would vary with the size of the storm. A detention basin with a 1 m depth and a surface area equal to or greater than 1% of the watershed area has been suggested as a guideline for sizing basins (Scheuler, 1987). A basin of this size would provide adequate detention time to allow settling of about 85% of the particles. Striegl (1987) presented a case study of a small lake that received urban runoff. The investigation was to determine the efficiency of the lake in removing suspended sediments and metals. Trap efficiency for suspended sediment ranged from 91 to 95%, an amount higher than expected based on Stokes Law. Rausch and Schreiber (1981) conducted a three-year study of a small flood detention reservoir in central Missouri. The detention basin area was equal to 0.6% of the drainage area and retained 85% of the sediments. Detention basins constructed in two major arms of Green Valley Lake, Iowa, were effective in reducing sediment and nutrient loads to the lake from runoff in an agricultural watershed (U.S. EPA, 1995).

In-stream treatment has been used before, primarily in drinking water reservoirs. The costs of such treatment were considered to be high, but were justified because of savings in water treatment. In the Wahnbach Reservoir, Germany, water quality had deteriorated so much that taste and odor problems were common and the mass algae blooms interfered with the water treatment process of flocculation and disinfection. A phosphorus elimination plant was built at the head of the reservoir, where about 90% of the phosphorus entered. The treatment plant is very effective at removing phosphorus, turbidity, algae and bacteria; removal rates exceeded 90% for these parameters (Bernhardt, 1980). Hayes et al. (1984) describe measures taken at the Foxcote Reservoir in Great Britain, including an application of ferric sulfate to the reservoir and chemical treatment of the pumped river inputs. The treatment effectively reduced the phosphorus concentrations in inflow, but phosphorus recycling from bottom sediments was sufficient to cause algal blooms. However, after three years of treatment, the water quality improved. Cooke and Carlson (1986) described an unsuccessful effort to use this technique in the restoration effort for Lake Rockwell, a eutrophic water supply reservoir that receives 90% of its flow from the Cuyahoga River. Although treatment removed over half of the soluble reactive phosphate, total phosphorus loading to the reservoir was not reduced. Aluminum floc built up rapidly on the river bed and pH dropped to 4, resulting in mortality of stream invertebrates near the treatment site. This treatment alternative was abandoned, and river conditions returned to normal.

5.12 Summary

The management strategies that may be feasible for implementation in some reclaimed water reservoirs are summarized on Table 5-1. For each strategy, the problems addressed, effectiveness, and other considerations are summarized.

Table 5-1. Summary of Reservoir Management Strategies

Strategy	Problems Addressed	Effectiveness	Other Considerations
Destratification/ Artificial Circulation	Low DO, odors, phosphorus release from sediments, algae indirectly	Very effective for improving DO concentrations and reducing odors, inconsistent results for other problems.	Disrupts the thermocline. Could increase turbidity and algae. Possible noise problems with compressed air.
Hypolimnetic Aeration/ Oxygenation	Low DO, odors, phosphorus release form sediments	Very effective for improving DO concentrations and reducing odors, inconsistent results for phosphorus release from sediments.	Maintains the thermocline. Commercial oxygen alleviates noise concerns.
Wetlands Treatment	Nitrogen loading from reclaimed water, algae indirectly	Very effective at removing nitrogen from reclaimed water. May require nitrogen removal at the treatment plant first to achieve desired TIN concentration. Improvements in water quality will be seen only if nitrogen is the limiting nutrient.	Some wetlands must have a constant supply of water; all require maintenance. To ensure effectiveness, other sources of nitrogen should be controlled.
Copper Treatment	Algae	Effective on a short-term basis. Does not address source of problem.	Temporary measure. May increase copper concentrations in sediments.
Phosphorus Removal	Phosphorus loading from reclaimed water, algae indirectly	Very effective at removing phosphorus from reclaimed water. Total phosphorus concentrations of < 0.1 mg/L likely required to affect algae concentrations. Improvements in water quality will be seen only if phosphorus is the limiting nutrient.	Relatively high costs. To ensure effectiveness, other internal and external sources of phosphorus should be controlled.
Nitrogen Removal	Nitrogen loading from reclaimed water, algae indirectly	Effective at removing nitrogen from reclaimed water. Wetlands treatment may also be needed to achieve desired TIN concentration. Improvements in water quality will be seen only if nitrogen is the limiting nutrient.	Relatively high costs. To ensure effectiveness, other sources of nitrogen should be controlled.
Dredging	Phosphorus release from sediments, low DO	Effective method to control internal source of phosphorus and sediment oxygen demand. Only effective for reclaimed water reservoirs if nutrient concentrations in reclaimed water are reduced.	Relatively high costs, but long-term effectiveness. Sediment testing required. Disposal may be a problem.
Alum Treatment	Phosphorus release from sediments, low DO	Can be an effective method to control internal source of phosphorus, depending on reservoir operations and nutrient concentrations in reclaimed water.	Additional treatments would be required every few years.
Watershed Management	Nutrient and sediment loads from the watershed	Can be an effective method to reduce nutrient and sediment loads from the watershed. Improvements in water quality will be seen only if other sources of nutrients (internal and in reclaimed water) are also controlled. A wide variety of BMPs could be implemented as part of a watershed management plan.	Plans should be specific to the individual watershed.

CHAPTER 6.0

Reservoir Water Quality Models

6.1 Overview

Most computer models that may be applicable to open reclaimed water reservoirs were developed for natural lakes or reservoirs. Many focus on eutrophication, which is also an issue for reclaimed water reservoirs. Available models include empirical, process, steady-state, and dynamic types. Empirical models predict water quality using relationships developed from correlations observed in field measurements. Process, or deterministic, models simulate the physical and chemical process occurring in the lake. Rate coefficients used in these analyses are usually derived from laboratory experiments or field measurements. The values of some parameters typical in reclaimed water are well outside the range normally observed in lakes and reservoirs. So while it is possible to use these models for reclaimed water, it is important to verify the model results for conditions beyond the values used in model development.

Water quality models were evaluated in three areas:

- Their ability to simulate conditions present in reclaimed water reservoirs.
- Their effectiveness in addressing the issues of key water quality parameters, water quality criteria, and management strategies.
- Model requirements and limitations.

Because of the range of facilities, operations, and end uses of reclaimed water, there is no one perfect model for open reclaimed water reservoirs. However, with an understanding of model capabilities in the three areas identified above, water utilities storing reclaimed water can decide which model would best suit their needs.

Capabilities, requirements, and limitations of each model are discussed in the sections below. Capabilities include factors such as the ability to simulate one, two, or three dimensions; interactions at the sediment-water interface; the types of parameters included in the model; and temporal variability. In preceding chapters, water quality criteria, important parameters and management strategies were discussed. The ability of a model to address these issues is an important element of model capabilities. Model requirements include computer operating system requirements; data requirements for calibration and maintenance of the model; and requirements of the modeler in terms of computer and limnological expertise. Limitations of the model define situations that can not be simulated by the model.

6.2 EUTROMOD

6.2.1 General Model Description and Applications

EUTROMOD is a spreadsheet based eutrophication model. It is an empirical steady-state model used to predict lake response to external nutrient loadings under well-mixed conditions. It can be used to predict lakewide growing season average conditions based on annual average nutrient loadings, and to guide decisions on nutrient control programs for reservoirs.

6.2.2 Model Processes

The EUTROMOD model estimates nutrient loading from the watershed and in-lake concentrations of nutrients and trophic state indicators such as chlorophyll *a* concentrations, Secchi disk depth, and the probability that blue-green algae are the predominant algae. It uses data obtained in the U.S. Environmental Protection Agency (U.S. EPA) national eutrophication study, trophic state models, and reservoir response models to calculate in-lake nutrient concentrations and other trophic state indicators based on nutrient loading from the watershed.

6.2.3 Model Requirements

EUTROMOD is personal computer (PC) compatible and is available from the North American Lakes Management Society (NALMS). Input requirements are:

- Watershed information such as drainage areas, land uses, and soils information.
- Climate data including precipitation and evaporation.
- Reservoir surface area and depth.

EUTROMOD is a relatively simple model to use, and limited computer expertise is required. Some limnological knowledge is needed to interpret results.

6.2.4 Limitations

EUTROMOD is a steady-state model that predicts lakewide averages. Therefore, it is not suitable for reservoirs where spatial and temporal variations are important. The model is limited to investigations of nutrient loading and eutrophication response; it does not simulate sediments or temperature.

6.2.5 Effectiveness for Reclaimed Water Reservoirs

The nutrient concentrations typically found in open reclaimed water reservoirs are well outside of the range of the concentrations used to develop the model. Therefore, care must be used in applying this model to reservoirs with very high nutrient concentrations. The EUTROMOD model could be a useful tool to examine the effects of decreases in phosphorus loading from the watershed. Most in-lake management strategies could not be represented directly in the model, but it could simulate changes in external loadings.

6.3 PHOSMOD

6.3.1 General Model Description and Applications

PHOSMOD is a mass balance model developed to predict the long-term response of a lake to changes in its phosphorus loading. It is based on a consideration of external and internal phosphorus loadings from sediments to calculate in-lake phosphorus and hypolimnetic oxygen concentrations. PHOSMOD is a steady state model that can be used to assess long-term trends only.

6.3.2 Model Processes

The development of the PHOSMOD model along with the specific equations it uses are described by Chapra and Canale (1991). The model is designed to compute total in-reservoir phosphorus concentrations and hypolimnetic oxygen concentrations, considering sediment water interactions. The lake is segmented into two layers: water and sediment. The phosphorus budget accounts for external loading and recycling of phosphorus from the sediments. The sediment layer can gain phosphorus through sedimentation and loses phosphorus by recycling and burial. The recycling rate of phosphorus from the sediments is dependent on the concentrations of phosphorus in the sediments and on the hypolimnetic oxygen concentration.

6.3.3 Model Requirements

PHOSMOD is PC compatible and is distributed by NALMS. Data requirements include physical characteristics of the reservoir, initial in-lake phosphorus and dissolved oxygen (DO) concentrations, sediment parameters, settling and burial rates for sediments, information on inflow rates and concentrations, and information on reservoir stratification. The model relies on site-specific data for calibration. PHOSMOD is relatively simple to use, and limited computer expertise is required. Some limnological training is needed to develop the model, to perform calibration, and to interpret results.

6.3.4 Limitations

PHOSMOD was developed for the specific purpose of evaluating reservoir responses to changes in phosphorus loading. It does not simulate short-term variations in reservoir water quality or spatial variations across the reservoir. It does not consider the effects of nitrogen loading, nor does it estimate chlorophyll *a* concentrations.

6.3.5 Effectiveness for Reclaimed Water Reservoirs

The phosphorus concentrations typically found in open reclaimed water reservoirs are well outside of the range of the concentrations used to develop the PHOSMOD model. Therefore, care must be used in applying this model to reservoirs with very high nutrient concentrations. However, if data are available to calibrate the model, it may provide useful results. It could be used to simulate the effects of reduced phosphorus loads that would occur after advanced treatment for phosphorus removal from the reclaimed water. Other management strategies could not be simulated directly by the model. PHOSMOD is a relatively simple model that could be a useful tool to aid in the management of open reclaimed water reservoirs if calibration data were available.

6.4 BATHTUB

6.4.1 General Model Description and Applications

BATHTUB is an empirical eutrophication model for both morphometrically simple and complex reservoirs. The program performs water and nutrient balance calculations in a steady-state, spatially segmented hydraulic network, which accounts for advective transport, diffusive transport, and nutrient sedimentation. Eutrophication-related water quality conditions are predicted using empirical relationships developed and tested previously for reservoir applications (Walker, 1987). BATHTUB can also be configured for simultaneous application to collections or networks of reservoirs to provide regional perspectives on reservoir water quality, controlling factors, and model performance. The most recent version of BATHTUB is described by Walker (1996).

BATHTUB can be used as either a diagnostic or predictive tool. Typical applications include:

- Development of water and nutrient balances;
- identification of factors controlling algal production;
- assessing impacts of changes in water and/or nutrient loads;
- assessing impacts of changes in mean pool level or morphometry;
- estimating nutrient loads consistent with given water quality management objectives.

6.4.2 Model Processes

The core of BATHTUB consists of the following processes:

- Water balance;
- nutrient balance (phosphorus and nitrogen);
- eutrophication response (chlorophyll *a*, Secchi disk depth);
- hypolimnetic and metalimnetic oxygen depletion.

The water balance yields a system of simultaneous linear equations that are solved to estimate the advective outflow from each model segment. Nutrient mass balances are expressed as a system of simultaneous nonlinear equations, which are solved iteratively. A second-order decay model is used because it is generally the most applicable formulation for representing phosphorous and nitrogen sedimentation in reservoirs, but alternative equations for representing nutrient sedimentation are also available. Total phosphorous and nitrogen concentrations resulting from mass balance are subsequently input to the model network to estimate eutrophication response in each segment. Chlorophyll *a* can be calculated using several options with different limiting factors, such as phosphorus, nitrogen, light, and flushing rate. Secchi disk depth and hypolimnetic oxygen depletion can also be calculated. The model can be calibrated to modify estimated nutrient concentrations, chlorophyll *a* concentration, Secchi depths, oxygen depletion rates, and dispersion coefficients.

BATHTUB provides optional output formats for various purposes, such as listing input conditions; summarizing hydraulic and dispersion calculations; summarizing water and mass balance; comparing observed and predicted water quality conditions; presenting observed values, estimated values and error ration in tables or graphics; and providing a sensitivity analysis.

6.4.3 Model Requirements

BATHTUB is PC compatible and is available on the Internet from the U.S. Army Corps of Engineers. Input requirements include:

- Reservoir characteristics such as volume, area, length, and thermocline depth.
- Water and nutrient loads from point sources and tributaries. Nonpoint sources can be estimated by the model if watershed information is provided.
- In-lake water quality data for calibration.

BATHTUB is a relatively simple model to use; it is menu-driven, and limited computer expertise is required. Some limnological training is needed to specify appropriate model options, perform calibration, and interpret results.

6.4.4 Limitations

BATHTUB is a steady-state empirical eutrophication model and therefore can not simulate dynamic conditions such as rapid changes in inflows or outflows or changes in pool elevations. Within each segment, the water quality is assumed to be uniform. Therefore, the model does not simulate longitudinal, lateral, and vertical distributions of water quality within a segment. The model is limited to investigations of nutrient loading and eutrophication response; it does not simulate sediments or temperature.

6.4.5 Effectiveness for Reclaimed Water Reservoirs

BATHTUB is an empirical model that relates eutrophication symptoms to external nutrient loads, hydrology, and reservoir morphometry, using statistical models derived from a representative cross section of reservoirs. Biochemical processes are not simulated explicitly, and simplified steady-state hydrodynamic representations are used. The model, which deals with temporally averaged conditions using a yearly or seasonal average, has been used widely over a number of years. For example, Kennedy (1995) applied the model to four reservoirs experiencing extensive development, and Ashby and Kennedy (1999) used it to predict the water quality of a proposed new reservoir. Although BATHTUB is a fairly simple model, its results have been shown to be as accurate as those of more complex models in some situations (Ernst, Frossard, and Mancini, 1994).

The nutrient concentrations typically found in open reclaimed water reservoirs are well outside of the range of the concentrations used to develop the model. Therefore, care must be used in applying this model to reservoirs with very high nutrient concentrations. Most in-lake management strategies could not be represented directly, but changes in external loadings could be simulated. For example, the model could be used to simulate the effects of reduced nutrient loads that would occur after advanced treatment for phosphorus or nitrogen removal from the reclaimed water. The Irvine Ranch Water District (IRWD) used the BATHTUB model for this purpose when the district was evaluating various reservoir and watershed management strategies for its two reclaimed water storage reservoirs (IRWD and Black & Veatch, 1994).

The BATHTUB model could be a useful tool to aid in the management of reclaimed water storage reservoirs. It is a relatively simple model that could provide meaningful results if calibration data were available.

6.5 CE-QUAL-R1

6.5.1 General Model Description and Applications

CE-QUAL-R1 is a one-dimensional reservoir water quality model. The reservoir is represented as a series of horizontal layers. Water within each layer is assumed to be well-mixed. The differences in densities of inflows, outflows, and water within each layer are an important element of the model. Inflows are directed to a reservoir layer based on density. The model simulates surface flows and interflows. Water quality within each layer is also simulated.

6.5.2 Model Processes

Model processes are more fully described in U.S. Army Corps of Engineers (2000). The model is structured as a series of horizontal layers whose thicknesses depend on the balance of inflows and outflows to maintain a mass balance. The thermal analysis portion of CE-QUAL-R1 is conducted using the model CE-THERM-R1. Vertical transport of heat and other constituents occurs through entrainment and turbulent diffusion. The model simulates interaction among three major components:

- Physical properties including flow, temperature, and total dissolved solids (TDS);
- chemical properties including nitrogen, phosphorus, carbon, and silica;
- biological groups including three algal groups, fish, and macrophytes.

Interactions are simulated for aerobic and anerobic conditions.

6.5.3 Model Requirements

CE-QUAL-R1 is PC compatible and is available from the U.S. Army Corps of Engineers. CE-QUAL-R1 includes several utilities, such as preprocessors, graphics utilities, and statistics. Input requirements include:

- Physical characteristics of the reservoir such as area, depth, and temperature profiles.
- Inflow information including rates, temperature, TDS, and concentrations of water quality constituents.
- In-reservoir concentrations of water quality parameters.

The User's Manual for the model (U.S. Army Corps of Engineers, 1982) provides information on how to select coefficients and rate constants needed to run the model and provides guidance on calibration procedures. More technical expertise, both modeling experience and limnological understanding, is required to run this model than to run the simpler models.

6.5.4 Limitations

CE-QUAL-R1 is a one-dimensional model. A major feature of the model is to account for density differences between inflow and reservoir water and to simulate interflows. Variations with depth are simulated, but longitudinal and lateral differences are not represented.

6.5.5 Effectiveness for Reclaimed Water Reservoirs

CE-QUAL-R1 is capable of simulating most of the parameters of interest for open reclaimed water storage reservoirs. It simulates stratification and DO depletion in the hypolimnion. This model could be an effective tool in some situations. Sufficient data for calibration would be required to obtain meaningful results.

6.6 CE-QUAL-W2

6.6.1 General Model Description and Applications

CE-QUAL-W2 is a two-dimensional hydrodynamic and water quality model. It can be applied to most water body types and can be used in one- or two-dimensional simulations. It is assumed that lateral variations across a lake or reservoir can be ignored. Because of this assumption, the model is best suited to reservoirs that are relatively long and narrow. However, multiple branches of a reservoir can be simulated. The model has been applied to over 100 systems since its development (Cole and Tillman, 1997). Its hydrodynamic module predicts water surface elevations and velocities in the horizontal and vertical directions. The hydrodynamic module is directly linked to a water quality module that predicts time-varying concentrations of water quality parameters.

6.6.2 Model Processes

Model processes are described in detail in the User's Manual for the model (Cole and Buchak, 1995). CE-QUAL-W2 uses a finite-difference method to solve the laterally averaged equation of motion. The reservoir is represented by a grid consisting of a series of vertical segments and horizontal layers. The hydrodynamic calculations consider the effects of variable water density caused by differences in temperature and TDS. The model simulates the interactions of many biological and chemical factors that affect water quality. Specific processes simulated in the model include:

- Temperature and salinity;
- the DO-carbon balance;
- the nitrogen cycle;
- the phosphorus cycle;
- the silicon cycle;
- phytoplankton;
- bacteria;
- first-order decay.

Up to 21 water quality constituents can be included in the model, in addition to temperature and velocity. The water quality constituents are separated into four levels of complexity, which allows for flexibility in applying the model to different situations.

6.6.3 Model Requirements

The CE-QUAL-W2 model is PC compatible and is available from the U.S. Army Corps of Engineers. Input data requirements include:

- Bathymetry of the reservoir sufficient to define the grid system.
- Initial conditions.
- Inflow rates and concentrations of water quality constituents, and outflow rates.
- Water surface elevations.
- Kinetic rate coefficients.

In-reservoir concentrations of water quality parameters at several locations are needed to calibrate the model. Model application and calibration are complex tasks that require a relatively high level of expertise for the modeler. Proper use of the model requires knowledge of modeling procedures, hydrodynamics, aquatic biology, aquatic chemistry, and statistics.

6.6.4 Limitations

CE-QUAL-W2 is a two-dimensional (horizontal and vertical) hydrodynamic and water quality model. Because lateral variations are ignored, the model is best suited to relatively long and narrow reservoirs. The model has one algal compartment and does not include zooplankton. Sediment interactions are represented by a simplified expression. A new version of the model is under development to improve some of these limitations.

6.6.5 Effectiveness for Reclaimed Water Reservoirs

CE-QUAL-W2 is a state-of-the-art, two-dimensional, hydrodynamic and water quality model that has been applied successfully to many lakes and reservoirs. Barnese and Bohannon (1994) discussed how it was used to describe the distribution of phytoplankton and nutrients in Taylorsville Lake in Kentucky. Cole and Tillman (1997) developed a model application for the Louisville District of the Corps of Engineers to evaluate changes in water quality due to environmental changes or changes in operating procedures of the district's reservoirs.

The CE-QUAL-W2 model is capable of simulating most of the processes and parameters important in reclaimed water reservoirs, including stratification, nutrient and DO cycles, and algae growth. This model uses detailed equations to simulate water quality kinetics and predict concentrations of important water quality parameters. If sufficient input and calibration data were available, the results of this model would most likely be more accurate than the simpler models discussed above. The model would be capable of simulating changes in nutrient inputs. However, the data and user requirements may be too extensive for most utility managers whose systems include open reclaimed water reservoirs.

6.7 RMA2 and RMA4

6.7.1 General Model Description and Applications

The Surfacewater Modeling System (SMS) is a comprehensive two-dimensional surface water modeling package that contains many individual models. Two of these, RMA2 and RMA4, can be used to model water quality in reservoirs. RMA2 is a two-dimensional, depth averaged, hydrodynamic numerical model that calculates surface water elevations and velocities in two-dimensional flow fields. RMA4 is a water quality transport model applicable to situations where variations in concentrations with depth can be ignored. It can be used to evaluate the mixing and transport of soluble substances or substances that can be represented by first-order decay. More detailed descriptions of these models are provided by BOSS International (2001a, b).

6.7.2 Model Processes

RMA2 can be used as either a steady state or dynamic model to predict water velocities and define the flow field. RMA2 computes a solution to the Navier-Stokes equations of motion using a finite element method. RMA4 uses the results produced by the RMA2 model. RMA4 computes concentrations for up to six water quality constituents as conservative or non-conservative using a first-order decay. From a set of initial conditions, RMA4 calculates concentrations using an advection-diffusion equation. It does not simulate biological or chemical processes, nor does it simulate eutrophication.

6.7.3 Model Requirements

The RMA2 and RMA4 models are part of the SMS software package. These models are provided by BOSS International and must be purchased. Input requirements for the RMA4 model include:

- Basic physical characteristics of the reservoir.
- Velocity fields (generated using RMA2).
- Initial conditions, boundary conditions and pollutant loads.

Input data requirements to specify initial and boundary conditions can be fairly extensive for a large system. Also, modeling experience is necessary to develop, calibrate, and run the model.

6.7.4 Limitations

The major limitation of this model is that it does not simulate the water quality constituents of primary concern in open reclaimed water reservoirs: nutrient cycles and reservoir response to nutrient loading. Additional limitations include the two-dimensional nature (depth averaged) nature of the model, data and user requirements, and costs.

6.7.5 Effectiveness for Reclaimed Water Reservoirs

The processes and parameters most important in reclaimed water reservoirs include stratification, nutrient and DO cycles, algae growth, and sediment-water interactions. The RMA4 model simulates the mixing and transport of primarily conservative substances. This model would not be very effective for reclaimed water storage reservoirs because it does not include the important processes and parameters.

6.8 MIKE 11 RESERVOIR

6.8.1 General Model Description and Applications

MIKE 11 is a professional engineering software package for the simulation of flows, water quality, and sediment transport in estuaries, rivers, reservoirs, and other water bodies. MIKE 11 is a suite of one-dimensional models based on an integrated modular structure with a variety of basic modules and add-on modules, each simulating certain phenomena in river systems.

MIKE 11 RESERVOIR was developed within the general software frame of MIKE 11. Different from MIKE 11, the core of the MIKE 11 RESERVOIR model is a two-dimentional vertical hydrodynamic flow model. It is designed to describe the hydrodynamic, transport, and water quality conditions of reservoirs using a multi-layered approach. The model can be used to study impacts of catchment inflow and operational strategies on the physical and biological processes within the reservoir, including testing and optimizing interventions and management solutions to alleviate existing problems in reservoirs and designing and assessing environmental impact of planned reservoirs. A more complete model description can be found in Danish Hydraulic Institute (DHI) (2000a).

6.8.2 Model Processes

MIKE 11 has several basic modules to describe a number of processes including:

- Advection and dispersion of conservative and non-conservative materials;
- cohesive sediment transport including erosion and deposition;
- salinity;
- BOD-DO;
- reaeration;
- bacteria;
- nitrogen cycle;
- phosphorous cycle;
- carbon cycle;
- phytoplankton and zooplankton cycle;
- benthic exchange;
- adsorption and desorption;
- resuspension and settling;
- diffusive transport of dissolved metal.

MIKE 11 RESERVOIR includes the following processes:

- Heating/cooling and evaporation processes;
- selective withdrawal (outflows) and inflows;
- siltation;
- reservoir flushing (dilution);
- eutrophication (plankton, nutrient, and oxygen dynamics);
- vertical oxygen profiles;
- ammonia concentration.

MIKE 11 RESERVOIR solves the laterally integrated Navier Stokes equation of motion combined with the equation for conservation of mass, salinity and temperature. The water quality module of MIKE 11 RESERVOIR can simulate vertical profiles of phytoplankton, nutrient, and DO depending on light conditions, water temperature, vertical mixing, and external and internal nutrient loads. The water quality module includes 12 state variables representing phytoplankton, zooplankton, detritus, inorganic nutrients, and DO.

Outputs from the model include time-series variables at each grid point. The Windows User Interface can make graphical presentations including time series plots, comparison of measured/simulated and simulated/simulated time series, animation of flow, etc.

6.8.3 Model Requirements

MIKE 11 and MIKE 11 RESERVOIR are available for Windows 95/98 and Windows NT platforms. These models are provided by the DHI and must be purchased. Costs depend on the specific modules included in the software package.

Input requirements include specification of boundary conditions such as topography and time series of inflows and outflows. Initial concentrations and boundary conditions must be supplied for each constituent. In addition, process constants, temperature coefficients, and sedimentation parameters must be specified. Although many model parameters are required, especially for the eutrophication module, default values are provided for many parameters and can be used if site-specific data are not available. Users need to have good understanding of the MIKE 11 module structures to make full use of the model. Also, significant modeling and limnological experience is necessary to develop, calibrate, and run the model.

6.8.4 Limitations

MIKE 11 RESERVOIR is a two-dimensional vertical hydrodynamic flow and water quality model. The model does not simulate cross-channel variations in water quality. Other limitations for use of this model are the data and user requirements and the cost of purchasing the model.

6.8.5 Effectiveness for Reclaimed Water Reservoirs

MIKE 11 RESERVOIR can predict dynamic flow conditions in stratified lakes and reservoirs with inflow and withdrawal assuming laterally averaged conditions. Water quality in the reservoir is described fully through BOD-DO, nutrients cycles, and plankton cycles, including exchange with bed sediment and temperature coefficients. The model can simulate phenomena such as algae growth, oxygen depletion/reaeration, considering nitrogen, phosphorus, light limits, and temperature effect. Bacteria decay can also be simulated. In addition, several of the management strategies could be simulated. If sufficient data are available and the model can be calibrated successfully, it could be a useful tool to help manage open reclaimed water reservoirs.

6.9 MIKE 21

6.9.1 General Model Description and Applications

MIKE 21 is a software package for two-dimensional, vertically averaged, dynamic flows. MIKE 21 is applicable to the simulation of flow and water quality in lakes, reservoirs, estuaries, and bays where stratification can be neglected. The model simulates the spreading of a dissolved or suspended substance under the influence of the fluid transport and associated natural dispersion process. The substance may be a conservative or linearly decayed pollutant, such as salt, heat, and dissolved/suspended contaminants. The water quality component of the model simulates the resulting concentrations of bacteria, oxygen depletion, concentration of nutrients, and chemical substances. The eutrophication component of the model is capable of simulating the annual cycle of phytoplankton and nutrients and the spatial variations that occur. The processes describing the variations of the components in time and space depend on external factors such as salinity, water temperature, light influx, and pollution loads.

6.9.2 Model Processes

The MIKE 21 model is described in more detail in DHI (2000b). The primary processes simulated in the model include:

- Advection and dispersion transport of dissolved or suspended substance (salt, heat, coliform bacteria, heavy metals and xenobiotic compounds);
- bacteria;
- oxygen conditions affected by BOD and other oxygen consuming substances;
- reaeration;
- eutrophication (algae blooming related to nutrient level);
- nutrient cycling;
- decay of chemical substances;
- benthic vegetation;
- adsorption/desorption of metals to suspended matter;
- sedimentation and resuspension of particulate metal;
- diffusive transport of dissolved metal at the sediment/water interface.

The water quality component of the model solves a system of differential equations describing the physical, chemical, and biological interactions involved in the survival of bacteria, oxygen conditions, and excess levels of nutrients. The variables modeled are BOD, ammonia, nitrate, DO, phosphorous, fecal coliforms, total coliforms, and one or more user defined pollutants. The eutrophication component of the model simulates phytoplankton, chlorophyll *a*, zooplankton, detritus, inorganic nitrogen, inorganic phosphorus, DO, and benthic vegetation carbon.

The output of the model includes the concentrations of each constituent varying with time and space. It provides a user-friendly interface based on a fully Windows-integrated Graphical User Interface, which facilitates the application of the system, including data preparation, analysis of simulation results, and graphical presentation.

6.9.3 Model Requirements

MIKE 21, available for the Windows 95/98 platforms, is provided by DHI, and must be purchased. The costs depend on the specific modules included in the software package.

Input requirements include specification of:

- Basic model parameters, including the model grid size and extent, time step and length of simulation, type of output required, and its frequency.
- Bathymetry and hydrodynamic inputs.
- Boundary and initial conditions providing compound concentrations, water temperature, light intensity, and salinity.
- Constituents parameters information, including process constants, temperature coefficient, and sedimentation rates.
- Source/sink discharge magnitude and compound concentrations.

Users need to have a good understanding of the MIKE 21 modules to make full use of the model. Also, modeling experience is necessary to calibrate the dispersion coefficients and decay coefficients before carrying out simulations.

6.9.4 Limitations

In hydrodynamic modeling of MIKE 21, vertically averaged conditions are assumed. For shallow reservoirs, averaged conditions may be a valid assumption, but for deep reservoirs that stratify the model may not be applicable. Other limitations are the data and user requirements and the cost of purchasing the model.

6.9.5 Effectiveness for Reclaimed Water Reservoirs

MIKE 21 can predict dynamic flow conditions in non-stratified lakes and reservoirs. Water quality in the reservoir is described fully through BOD-DO, nutrients cycles, and plankton cycles, including exchange with bed sediment and temperature coefficients. The model can reproduce the phenomena in a non-stratified reservoir/lake such as algae growth, oxygen depletion/reaeration, considering nitrogen, phosphorus, light limits, and temperature effect. Bacteria decay can also be simulated. If sufficient data are available and the model can be calibrated successfully, it could be a useful tool to help manage shallow open reclaimed water reservoirs.

6.10 CE-QUAL-ICM

6.10.1 General Model Description and Applications

The CE-QUAL-ICM model is a three-dimensional eutrophication model. It was developed originally as part of the modeling package for the three-dimensional model of Chesapeake Bay (Cerco and Cole, 1993). It can be used to predict time-varying concentrations of water quality parameters. To apply the model, the study area must be represented by a grid of discrete volumes or cells. The grid can be arranged in one-, two-, or three-dimensional configurations. The model does not calculate hydrodynamics. Flow, volumes, and diffusion coefficients must be calculated using an external hydrodynamic model, developed into a hydrodynamics file, and read into the model. Twenty-two variables, including physical properties, nutrients, DO, and algae, are simulated by the model. Detailed information on the model features, capabilities, and requirements are provided in the User's Manual (Cerco and Cole, 1995).

6.10.2 Model Processes

Interactions between the 22 state variables are simulated using a large set of partial-differential equations. The conservation of mass equations are solved for each parameter in each grid of the model using a finite-difference method. The model has full capabilities to simulate state-of-the-art water quality kinetics. The specific processes simulated by the CE-QUAL-ICM model are:

- Temperature and salinity;
- DO and carbon cycle;
- nitrogen cycle;
- phosphorus cycle;
- silica cycle;
- three forms of algae;
- sediment-water oxygen and nutrient fluxes.

6.10.3 Model Requirements

The CE-QUAL-ICM model is PC compatible and is available from the U.S. Army Corps of Engineers. The CE-QUAL-ICM model has extensive data requirements for both initial inputs and calibration. Input data requirements include geometric data to specify the computational grid and about 140 parameters to solve the conservation of mass equations. Initial and boundary conditions must be specified. An extensive database would be required for calibration, because the model simulates an extensive number of properties and processes. Although the model is PC compatible, computational requirements can be large if many processes are simulated within a large study area. The model requires significant technical expertise of the user to specify initial parameters, calibrate the model, and interpret the results. Technical expertise in modeling, limnology, and aquatic biology would be required.

6.10.4 Limitations

The CE-QUAL-ICM model is a dynamic model, but does not compute the hydrodynamics, which must be supplied externally. In addition, although the model has significant capabilities to simulate water quality processes, the data requirements are extensive, and the users must have significant technical expertise.

6.10.5 Effectiveness for Reclaimed Water Reservoirs

The CE-QUAL-ICM model is capable of simulating the processes and parameters important in reclaimed water reservoirs, including stratification, nutrient and DO cycles, algae growth, and sediment-water interactions. This model uses detailed kinetics equations to simulate water quality and predict concentrations of important water quality parameters. If sufficient input and calibration data are available, the results of this model would most likely be more accurate than those of the simpler models discussed above. However, the data and user requirements may be too extensive for most managers of reclaimed water systems.

6.11 WASP5

6.11.1 General Model Description and Applications

WASP5 is a dynamic compartment modeling program for aquatic systems, including both the water column and the underlying benthos. It can be applied to most water bodies, such as ponds, streams, lakes, reservoirs, rivers, estuaries, and coastal waters, in one, two, and three dimensions, which can be structured by the modeler. The program can be used to simulate time-varying processes of advection, dispersion, point and diffuse mass loading, boundary exchange, and kinetics of various water quality constituents. It can be linked with hydrodynamic models such as DYNHYD5, RIVMOD, HYDRO3D, and others.

WASP5 includes two submodels: EUTRO5 for water quality/eutrophication (involving DO, BOD, nutrients and eutrophication) and TOXI5 for toxicants (involving organic chemicals, metals, and sediment). TOXI5 can also simulate substances with no kinetic interactions (i.e., conservative substances) by specifying no decay.

6.11.2 Model Processes

Complete documentation for the WASP5 model is provided by Ambrose, Wool, and Martin (1993). The processes simulated in the model include:

- Advection and dispersion soluble constituents;
- pore water advection and diffusion;
- transport of particulate pollutants by settling, resuspension, and sedimentation of solids;
- salinity;
- bacteria;
- temperature;
- evaporation/precipitation;
- sediment transport;
- equilibrium reversible sorption;
- BOD-DO;
- phytoplankton kinetics;
- phosphorus cycle;
- nitrogen cycle;
- benthic exchange;
- transport and fate of chemicals (simple and organic chemicals);
- daughter product.

The model network of WASP5 is a set of segments. Segments may be one of four types as specified by the user. The four types indicate the epilimnion (surface water), the hypolimnion (subsurface), the upper benthic layers, and the lower benthic layers. Transport within and between these types of segments is simulated by the model. Transport fields include advective flow and dispersive mixing in the water column, movement of pore water in the sediment bed, transport of particulate pollutants by the settling, resuspension, and sedimentation of solids, and evaporation or precipitation from or to surface water segments.

EUTRO5 and TOXI5 can be operated by the user at various levels of complexity to simulate some or all of these variables and interactions. EUTRO5 simulates up to eight state variables both in the water column and sediment bed. These variables can be considered as four interacting systems: phytoplankton kinetics, the phosphorus cycle, the nitrogen cycle, and the DO balance.

EUTRO5 reports a set of state variable concentrations, forcing functions, and process rates for every segment at the specified output time interval. Variable concentrations include DO, carbonaceous BOD, ultimate BOD, phytoplankton carbon and chlorophyll *a*, total nitrogen, ammonia, nitrate, organic nitrogen, total inorganic nitrogen (TIN), organic phosphorus, and inorganic phosphorus.

6.11.3 Model Requirements

WASP5 is PC compatible and available from the U.S. EPA on the Internet. Pre- and post-processors are available from the distributor. Input data requirements and the data required for calibration can be extensive. Input data requirements depend on the level of complexity defined by the user. Input requirements can be grouped as follows:

- Simulation and output control.
- Model segmentation (number of segments).
- Advective and dispersive transport.
- Boundary concentrations.
- Point and diffuse source waste loads.
- Kinetic parameters, constants, and time functions.
- Initial concentrations.
- Sediment transport data.
- Hydraulic data (from a coupled hydrodynamic model).

Because this model can be used as a one-, two-, or three-dimensional model, the level of complexity differs depending on the model configuration. Calibration of the model can be difficult. Tufford and McKeller (1999) tried to calibrate it using phytoplankton kinetic rates and constants obtained from field studies, but concluded that other processes, not included in the model, must be present in the lake. To use the full capabilities of the model, the users must have considerable technical expertise in both modeling and limnology.

6.11.4 Limitations

WASP5 must be coupled externally with a hydrodynamic model to simulate stratified lakes/reservoirs. There is a potential for instability or numerical dispersion in the user-specified computational network. Time steps are constrained within a specific range to maintain stability and minimize numerical dispersion.

6.11.5 Effectiveness for Reclaimed Water Reservoirs

The WASP model has been used widely in the United States for a variety of water quality investigations. Eutrophication (phytoplankton growth) and DO depletion/reaeration can be simulated both in water column and bed sediment. Temperature, light, and nutrients limits are taken into consideration in simulating eutrophication and DO. Bacteria decay is also simulated. Hernandez et al. (1997) used WASP5 to predict daily phytoplankton and nutrient concentrations in the Amadoria Reservoir in Spain. Lung and Larson (1995) used the model to evaluate several scenarios for reducing phosphorus loading to the upper Mississippi River and Lake Pepin.

The WASP5 model is capable of simulating the processes and parameters important in reclaimed water reservoirs, including stratification, nutrient and DO cycles, algae growth, and sediment-water interactions. If sufficient input and calibration data are available, the results of this model would most likely be more accurate than those of the simpler models discussed above. However, the data and user requirements may be too extensive for most reclaimed water utility managers.

6.12 MIKE 3

6.12.1 General Model Description and Applications

MIKE 3 is a generalized mathematical modeling system designed for a wide range of applications in areas such as oceanography, coastal regions, estuaries, and lakes. It is a fully three-dimensional hydrodynamic model. It can be applied to stratified flow, water pollution studies, environmental impact assessment studies, and sedimentation studies.

MIKE 3 is composed of a series of modules that simulate hydrodynamics, advection and dispersion, water quality, eutrophication, and sediment transport. The water quality module describes BOD-DO relationships, nutrients, and bacteria concentrations. The eutrophication module simulates algae growth and primary production. The mud transport module simulates transport of mud along with erosion and deposition of cohesive material.

6.12.2 Model Processes

The MIKE 3 model is described more fully by DHI (2000c). The primary processes simulated by the model include:

- Salinity;
- temperature;
- density Variations;
- turbulence;
- BOD-DO (reaeration);
- horizontal and vertical transport of oxygen;
- bacteria;
- nitrogen cycle;
- phosphorous cycle;

- carbon cycle;
- phytoplankton and zooplankton cycles;
- benthic vegetation;
- light attenuation;
- exceeding concentration;
- cohesive/non-cohesive sediment.

The water quality module solves the system of differential equations describing the physical, chemical, and biological interactions involved in the survival of bacteria, oxygen conditions, and excess levels of nutrients. The variables modeled are BOD, ammonia, nitrate, DO, phosphorous, and coliforms. The influences of solar radiation and water temperature are considered in the simultaneous computations of these variables. The eutrophication module simulates phytoplankton, chlorophyll *a*, zooplankton, detritus, inorganic nitrogen, inorganic phosphorus, DO, and benthic vegetation carbon.

The output from the model includes the concentrations of every constituent varying with time and space. MIKE 3 provides a user-friendly interface based on a fully Windows-integrated Graphical User Interface, which facilitates the application of the system, including data preparation, analysis of simulation results, and graphical presentation.

6.12.3 Model Requirements

MIKE 3 is available for the Windows 95/98 platforms. This model is provided by DHI and must be purchased. The costs depend on the specific modules included in the software package. Input requirements include specification of:

- Basic model parameters including the model grid size and extent, time step and length of simulation, and the type of output required and its frequency.
- Bathymetry and hydrodynamic input.
- Boundary and initial conditions providing compound concentrations, water temperature, light intensity, and salinity.
- Constituents parameter information including process constants, temperature coefficient, and sedimentation rates.
- Source/sink discharge magnitude and compound concentrations.

Users need to have good understanding of the MIKE 3 modules to make full use of the model. Also, modeling experience is necessary to calibrate the dispersion coefficients and decay coefficients before carrying out simulations. Overall, extensive computer and limnological expertise is required of the user.

6.12.4 Limitations

The primary limitations of the MIKE 3 model are the cost and the extensive data and user requirements.

6.12.5 Effectiveness for Reclaimed Water Reservoirs

MIKE 3 can predict dynamical flow conditions in stratified lakes and reservoirs. Water quality in the reservoir is described fully through BOD-DO relationships, nutrients cycles, and plankton cycles, including exchange with bed sediment and temperature coefficients. The model can simulate phenomena such as algal growth, oxygen depletion/reaeration, considering nitrogen, phosphorus, light limits, and temperature effect. Bacteria decay can also be simulated. In addition, several of the management strategies could be simulated. If sufficient data are available and the model can be calibrated successfully, it could be a useful tool to help manage open reclaimed water reservoirs.

6.13 CWR Computational Aquatic Ecosystem Dynamics Model (CAEDYM)

6.13.1 General Model Description and Applications

CAEDYM is an ecological model developed by the Centre for Water Research (CWR) at the University of Western Australia. The model is designed to link to a number of hydrodynamic models, or it can be used alone if spatial variability can be ignored. CAEDYM is designed to be linked to other CWR hydrodynamic models, including the Dynamic Reservoir Simulation Model (DYRESM) and the Estuary, Lake, and Coastal Ocean Model (ELCOM) or to other hydrodynamic models such as CE-QUAL-W2. DYRESM is a one-dimensional model used for predicting vertical density stratification in reservoirs. ELCOM is a three-dimensional hydrodynamic model applicable to lakes, reservoirs, and coastal areas. The CAEDYM model was designed primarily to investigate eutrophication problems and includes simulations of nutrient, phytoplankton, zooplankton, and DO. The model is also capable of simulating seven major biological groups, including phytoplankton, zooplankton, fish, macroalgae, invertebrates, seagrass, and jellyfish.

6.13.2 Model Processes

Detailed information on model processes is provided by Herzfeld and Hamilton (2000a). The CAEDYM model simulates nutrients, phytoplankton, zooplankton, DO, and other state variables using temporal and spatial scales selected by the user. An external hydrodynamic model must be used to provide information on the flow field. Specific processes included in the model are:

- Temperature and salinity;
- DO, BOD, and the carbon cycle;
- nitrogen cycle;
- phosphorus cycle;
- silica cycle;
- seven forms of phytoplankton (marine and freshwater);
- five classes of zooplankton;
- five other major biological groups;
- sediment-water oxygen and nutrient fluxes;
- resuspension of particulate nutrients.

6.13.3 Model Requirements

The User's Manual for the CAEDYM model (Herzfeld and Hamilton 2000b) describes fully the input requirements and the procedures needed to run the model. Input requirements include specification of:

- Basic model parameters including the model grid size and extent, time step and length of simulation, and the type of output required and its frequency.
- Definition of the flow field supplied by an external hydrodynamic model.
- Boundary and initial conditions for the parameters simulated.
- Process constants and rate coefficients.
- Sedimentation data.

The model has extensive input and user requirements. Many parameters need to be specified to run and calibrate this model. The modeler should have significant experience with modeling and understanding of limnological processes. At present, the graphic user interface for this model is still under development, so data input is more difficult than with most of the other models. CAEDYM and the hydrodynamic models DYRESM and ELCOM are distributed by the Centre for Water Research through a licensing agreement. Royalty fees may be required for some models.

6.13.4 Limitations

The CAEDYM model is a dynamic model, which must be linked to an external hydrodynamic model. In addition, although the model has significant capabilities to simulate water quality processes, the data requirements are extensive, and users must have significant technical expertise.

6.13.5 Effectiveness for Reclaimed Water Reservoirs

The CAEDYM model is capable of simulating the processes and parameters important in reclaimed water reservoirs including stratification, nutrient and DO cycles, algae growth, and sediment-water interactions. This model uses detailed sets of equations to simulate water quality kinetics and predict concentrations of important water quality parameters. If sufficient input and calibration data are available, the results of this model would most likely be more accurate than those of the simpler models discussed above. However, the data and user requirements may be too extensive for most managers of reclaimed water systems.

6.14 Summary

The water quality computer models available for open reclaimed water reservoirs are summarized in Table 6-1. For each water quality model listed, the type, parameters simulated, processes simulated, complexity, data and user requirements, as well as other factors are presented.

Table 6-1. Summary of Reservoir Model Characteristics and Capabilities

Model	Type	Parameters Simulated	Processes Simulated	Complexity (C) Data and User Requirements (R)	Other Factors
EUTROMOD	Steady-state, well-mixed basin	Nitrogen, phosphorus, chlorophyll *a*, Secchi disk depth, probability that blue-green algae will dominate	Dilution, empirical relationships between watershed loads and in-reservoir nutrient concentrations and eutrophication response	C: Simple R: Low	Can simulate reservoir response to reduced nutrient loads. Does not consider stratification or sediment interactions.
PHOSMOD	Steady-state, water and sediment layers, well-mixed basin	Phosphorus, hypolimnetic DO concentrations	Dilution, empirical relationships between phosphorus loads from watershed and sediments and in-reservoir phosphorus concentration	C: Simple R: Low	Can simulate reservoir response to reduced phosphorus loads. Does not consider stratification.
BATHTUB	Steady-state, segmented reservoirs, well-mixed segments	Nitrogen, phosphorus, chlorophyll *a*, Secchi disk depth, hypolimnetic oxygen depletion	Dilution, Empirical relationships between nutrient loads and in-reservoir nutrient concentrations and eutrophication response	C: Simple R: Low	Can simulate reservoir response to reduced nutrient loads.
CE-QUAL-R1	Dynamic, 1-D, vertical (horizontally averaged)	Temperature, TDS, nitrogen, phosphorus, algae, DO, carbon, silica, bacteria	Heat balance, dispersion; DO-nutrients-algal cycle, carbon cycle, selective withdrawals	C: Moderate R: Moderate	Simulates stratification and hypolimnetic DO depletion. Can simulate reservoir response to reduced nutrient loads.
CE-QUAL-W2	Dynamic, 2-D, vertical and longitudinal (laterally averaged)	Temperature, TDS, velocity, nitrogen, phosphorus, algae, DO, BOD, carbon, silica, bacteria	Dilution, advection, dispersion, heat balance, selective withdrawal, DO-nutrients-algal cycle, carbon cycle, particle fate, sediment interactions, first-order decay	C: Complex R: Extensive	Simulates stratification and hypolimnetic DO depletion. Can simulate reservoir response to reduced nutrient loads, other strategies indirectly.
RMA2 and RMA4	Dynamic, 2-D, horizontal (depth averaged)	Horizontal velocities, conservative substances, non-conservative (first-order decay) substances	Dilution, advection, dispersion, first-order decay	C: Moderate R: Moderate	Does not simulate nutrient cycles or eutrophication processes.

Table 6-1. Summary of Reservoir Model Characteristics and Capabilities (cont.)

Model	Type	Parameters Simulated	Processes Simulated	Complexity (C) Data and User Requirements (R)	Other Factors
MIKE 11 RESERVOIR	Dynamic, 2-D, vertical and longitudinal (laterally averaged)	Temperature, TDS, velocity, nitrogen, phosphorus, algae, DO, BOD, carbon, bacteria	Advection, dispersion, heat balance, selective withdrawal, reaeration, DO-nutrients-algal cycle, carbon cycle, particle fate, sediment interactions, resuspension, first-order decay	C: Moderate R: Moderate	Simulates stratification and hypolimnetic DO depletion. Can simulate reservoir response to reduced nutrient loads, other strategies indirectly.
MIKE 21	Dynamic, 2-D, horizontal (depth averaged)	Temperature, TDS, velocity, nitrogen, phosphorus, algae, DO, carbon, silica, bacteria	Advection, dispersion, selective withdrawal, resuspension, reaeration, DO-nutrients-algal cycle, carbon cycle, particle fate, sediment interactions, first-order decay	C: Moderate R: Moderate	Does not consider stratification or hypolimnetic DO depletion. Can simulate reservoir response to reduced nutrient loads.
CE-QUAL-ICM	Dynamic, 3-D, link to external hydrodynamic model	Temperature, TDS, nitrogen, phosphorus, algae, DO, BOD, carbon, silica, bacteria	DO-nutrients-algal cycle, carbon cycle, silica cycle, sediment interactions, first-order decay	C: Complex R: Very Extensive	Simulates stratification and hypolimnetic DO depletion. Can simulate reservoir response to reduced nutrient loads, other strategies indirectly.
WASP5	Dynamic, 3-D, link to external hydrodynamic model	Temperature, TDS, nitrogen, phosphorus, algae, DO, BOD, carbon, silica, bacteria	Dilution, advection, dispersion, heat balance, DO-nutrients-algal cycle, carbon cycle, particle fate, sediment interactions, sediment transport, first-order decay	C: Complex R: Extensive	Simulates stratification and hypolimnetic DO depletion. Can simulate reservoir response to reduced nutrient loads, other strategies indirectly.
MIKE 3	Dynamic, 3-D	Temperature, TDS, velocity, nitrogen, phosphorus, algae, DO, carbon, silica, bacteria	Dilution, advection, dispersion, heat balance, DO-nutrients-algal cycle, carbon cycle, particle fate, sediment interactions,first-order decay	C: Complex R: Extensive	Simulates stratification and hypolimnetic DO depletion. Can simulate reservoir response to reduced nutrient loads, other strategies indirectly.
CAEDYM	Dynamic, 3-D, link to external hydrodynamic model	Temperature, TDS, nitrogen, phosphorus, algae, DO, BOD, carbon, silica, bacteria	Dilution, advection, dispersion, heat balance, DO-nutrients-algal cycle, carbon cycle, particle fate, sediment interactions, resuspension, first-order decay	C: Complex R: Extensive	Simulates stratification and hypolimnetic DO depletion. Can simulate reservoir response to reduced nutrient loads, other strategies indirectly.

CHAPTER 7.0

MANAGEMENT TOOLS

7.1 Overview

In the preceding chapters, water quality criteria, important parameters, management strategies, and water quality models were discussed. The information previously presented in this report is synthesized in this chapter to assist utility managers in selecting useful models and reservoir management strategies. First, the water quality problems typically found in reclaimed water reservoirs and the management strategies that may be effective in addressing these problems are summarized. The water quality models presented in Chapter 6.0 have been organized into three groups representing different levels of complexity: steady-state empirical models, one- or two-dimensional dynamic process models, and three-dimensional dynamic process models. For each group, the general characteristics, advantages and disadvantages, and individual models are summarized. Finally, guidance on model selection is provided, to help relate the models to specific water quality problems.

7.2 Common Water Quality Problems and Management Strategies

The typical water quality problems encountered in open reclaimed water reservoirs are excess nutrients and algae, low dissolved oxygen (DO), high turbidity, odors and color. These problems do not occur independently, but are interrelated. Most open reservoirs for reclaimed water storage provide ideal conditions for algae growth: ample nutrients from the reclaimed water and sunlight. Runoff from the watershed and internal sources of nutrients may add to the total nutrient load of the reservoir. Nitrogen and phosphorus typically do not become depleted enough to limit growth; other factors such as self-shading or zooplankton grazing are probably the growth limiting factors. In reservoirs deep enough to stratify, the lower layers can become effectively cut off from the surface and its supply of oxygen during the summer. As algae die, sink, and decay, they consume oxygen. Oxygen depletion can lead to the formation of hydrogen sulfide (H_2S), one cause of odors. Oxygen depletion also greatly increases the rate of release of phosphorus to the overlying water column. This internal source of phosphorus remains in the hypolimnion until fall turnover when it mixes throughout the water column.

The primary cause of water quality problems in reclaimed water reservoirs is excessive nutrient concentrations. The two primary problems are:

- Low DO, which leads to odors, fish kills and release of nutrients, iron, and manganese from sediments.
- Excessive algae, which can also lead to odors, cause filter clogging in the distribution system, and produce unsightly surface mats or scums.

Management strategies are designed to either address the cause of the problems (i.e., the various sources of nutrients) or treat the water quality problems directly. Management strategies that can be used to address excessive nutrients, low DO and excessive algae are listed in Table \7-1. There are three sources of nutrients in open reclaimed water reservoirs: the reclaimed water, watershed runoff, and internal sources. Management strategies for each of these are listed in Table 7-1. The relative importance of each of the three sources will vary for each reservoir. However, if nutrient control is selected as the management strategy, all significant sources of nutrients should be controlled to ensure effectiveness. Controlling the nutrient loading into the reservoir may be difficult and expensive. Other management strategies that could be used to address directly other water quality problems are also listed. Aeration either through destratification or oxygenation has been a very effective way to increase DO in reservoirs and alleviate hydrogen sulfide odors. In some cases, it has also reduced algae concentrations. Copper treatment of reservoirs can be effective at reducing algal blooms, but this is a short-term solution only.

Detailed information on each of the management strategies listed in Table 7-1 was presented in Chapter 5.0, along with an approach that could be used to evaluate and select an appropriate strategy for specific reservoirs. A water quality model could be a useful tool to help evaluate management strategies. Models could be used to determine the nutrient reductions needed to achieve concentration goals for a reservoir or to evaluate other management strategies.

Table 7-1. Strategies to Solve Water Quality Problems

Problem: Excessive Nutrients	
Management Strategies Reclaimed Water: total phosphorus or total nitrogen removal Internal Sources: dredging, alum, aeration External Sources: watershed management	Issues Addresses low DO and algae problems. Should reduce nutrients from all significant sources.
Problem: Low Dissolved Oxygen	
Management Strategies (Direct Control Techniques) Destratification Aeration	Issues Very effective way to increase DO, May not affect algae growth
Problem: Excessive Algal Growth	
Management Strategies (Direct Control Techniques) Copper Treatment	Issues Short-term solution only

7.3 Simple Empirical Models

Three of the models discussed in Chapter 6, EUTROMOD, PHOSMOD, and BATHTUB, are steady-state empirical models. Eutrophication-related water quality conditions are predicted using empirical relationships previously developed and tested for reservoir applications. These models predict annual or seasonal averages for the entire lake or for lake segments.

EUTROMOD is a spreadsheet based eutrophication model. It is an empirical steady-state model that predicts lake response to external nutrient loadings under well-mixed conditions. It predicts lakewide, growing season average conditions based on annual average nutrient loadings. Therefore, it is not suitable for reservoirs where spatial and temporal variations are important. The model is limited to investigations of nutrient loading and eutrophication response; it does not simulate sediments or temperature.

PHOSMOD is a steady state model developed to predict the long-term response of a lake to changes in its phosphorus loading. The model considers external phosphorus loading and internal loading from sediments to calculate in-lake phosphorus and hypolimnetic oxygen concentrations. The effects of nitrogen loading, one not considered, nor does the model estimate chlorophyll *a* concentrations. The lake is segmented into two layers: water and sediment. The phosphorus budget accounts for external loading and recycling of phosphorus from the sediments.

BATHTUB is an empirical eutrophication model that performs water and nutrient balance calculations in a steady-state, spatially segmented hydraulic network. Within each segment, the water quality is assumed to be uniform. Therefore, the model does not simulate longitudinal, lateral, and vertical distributions of water quality within a segment. The model is limited to investigations of nutrient loading and eutrophication response; it does not simulate sediments or temperature.

While each of these models is unique, they have some similar characteristics. They are relatively simple to use, and limited computer expertise is required. Some limnological knowledge is needed for calibration and to interpret results. The models also require only a limited amount of data. However, the nutrient concentrations typically found in open reclaimed water reservoirs are well outside of the range of the concentrations used to develop each model. Therefore, care must be used in applying these models to reservoirs with very high nutrient concentrations. However, if the models are calibrated to an individual reservoir, they could be useful in some situations. Another limitation of these models is their inability to simulate spatial and temporal variability. They can not account for changes in inflow rates or concentrations. Most in-lake management strategies could not be represented directly by these simple models, but changes in external loadings could be simulated. These models could be useful tools to examine the effects of decreases in phosphorus loading to the reservoir either through reductions in reclaimed water concentrations or reductions in loadings from the watershed. With a minimal effort, they could be used to guide decisions on nutrient control programs for reservoirs.

7.4 One- and Two-Dimensional Dynamic Models

Five of the models presented in Chapter 6, CE-QUAL-R1, CE-QUAL-W2, MIKE 11 RESERVOIR, MIKE 21, and RMA2 and 4, fit into the category of one-or two-dimensional dynamic process models. These models can be used to predict water quality in reservoirs using equation sets to simulate the processes that occur in the reservoirs. Such processes include advection, dispersion, heat transfer, sedimentation, nutrient cycles, oxygen depletion, reaeration, and biological growth. The models provide information on spatial and temporal variability.

CE-QUAL-R1 is a one-dimensional reservoir water quality model. The reservoir is represented as a series of horizontal layers. Variations in water quality with depth are simulated, but longitudinal and lateral differences are not represented. Water within each layer is assumed to be well mixed. CE-QUAL-R1 is capable of simulating most of the parameters of interest for open reclaimed water reservoirs including stratification and DO depletion in the hypolimnion.

CE-QUAL-W2 is a two-dimensional hydrodynamic and water quality model. It is assumed that lateral variations across a lake or reservoir can be ignored, and, therefore, the model is best suited to reservoirs that are relatively long and narrow. The hydrodynamic module is linked directly to a water quality module that is used to predict time-varying concentrations of water quality parameters. The model simulates the interactions of many biological and chemical factors that affect water quality. Up to 21 water quality constituents can be included, in addition to temperature and velocity. Sediment interactions are represented by a simplified expression.

MIKE 11 RESERVOIR is a two-dimensional (vertical and longitudinal) hydrodynamic flow model. It does not simulate cross-channel variations in water quality. The water quality module of MIKE 11 RESERVOIR can simulate vertical profiles of phytoplankton, nutrient, and DO, depending on light conditions, water temperature, vertical mixing, and external and internal nutrient loads. The water quality module includes 12 state variables representing phytoplankton, zooplankton, detritus, inorganic nutrients, and DO.

MIKE 21 is a software package for two-dimensional, vertically averaged, dynamic flows. The water quality component of the model simulates concentrations of bacteria, oxygen depletion, nutrients, and chemical substances. The eutrophication component of the model is capable of simulating the annual cycle of phytoplankton and nutrients and the spatial variations that occur. In the hydrodynamic modeling of MIKE 21, vertically averaged conditions are assumed. For shallow reservoirs, averaged conditions may be a valid assumption, but for deep reservoirs that stratify, the model may not be applicable.

RMA2 is a two-dimensional, depth averaged, hydrodynamic numerical model that can be linked to RMA4, a water quality transport model. These inked models are applicable to situations where variations in concentrations with depth can be ignored. The model can be used to evaluate the mixing and transport of soluble substances or substances that can be represented by first-order decay such as bacteria. The model does not simulate biological or chemical processes, nor does it simulate eutrophication. However, it could be useful in reservoirs where other parameters such as toxics or bacteria are important.

MIKE 21 and RMA4 are depth averaged models and, therefore describe variability in horizontal directions; it is assumed that the water column is homogeneous in the vertical direction. The other three models describe changes in the vertical direction; CE-QUAL-W2 and MIKE 11 RESERVOIR also describe changes along the axis of the reservoir. The depth average models are incapable of simulating stratification and would therefore be most appropriate in shallow reservoirs.

More technical expertise, both modeling experience and limnological understanding, would be required to run these models than is required to run the simpler models. However, if sufficient input and calibration data are available, the results of the one- and two-dimensional models should be more accurate than the steady state empirical models. With the exception of RMA4, these models are capable of simulating most of the processes and parameters important in reclaimed water reservoirs, including stratification, nutrient and DO cycles, and algae growth. The models would be capable of simulating changes in nutrient inputs from reclaimed water, the watershed, and internal sources. In addition, some of them could be used to examine the effects of other management strategies, such as aeration systems to increase hypolimnetic DO, dredging, and alum treatment.

Sufficient data for calibration and to define inputs adequately would be required to obtain meaningful results. If sufficient data are available and these models can be calibrated successfully, they can be a useful tool to help manage reclaimed water storage reservoirs. The one-and two-dimensional models can address some of the issues that cannot be addressed by the simpler models, including spatial and temporal variability and low DO in the hypolimnion. Although these models would provide more detailed and possibly more accurate results, the data and user requirements may be too extensive for most managers of reclaimed water systems. However, decisions regarding major operational changes or large capital improvements may warrant the additional effort required by these models.

7.5 Three-Dimensional Dynamic Models

Four of the models presented in Chapter 6.0, CE-QUAL-ICM, WASP5, MIKE 3, and CWR Computational Aquatic Ecosystem Dynamics Model (CAEDYM), fit into the category of three-dimensional dynamic process models. These models can be used to predict water quality in reservoirs using equation sets that attempt to simulate the processes that occur in the reservoirs. These processes include advection, dispersion, heat transfer, sedimentation, nutrient cycles, oxygen depletion, reaeration, and biological growth. The models provide information on spatial and temporal variability.

The CE-QUAL-ICM model is a three-dimensional eutrophication model. The model can be used to predict time-varying concentrations of water quality parameters. The model grid can arranged in one-, two-, or three-dimensional configurations. The model does not calculate hydrodynamics, and therefore must be linked to an external hydrodynamic model. There are 22 state variables, including physical properties, nutrients, DO, and algae simulated by the model. The CE-QUAL-ICM model has extensive data requirements for both initial inputs and calibration.

WASP5 is a dynamic compartment modeling program for aquatic systems, including both the water column and the underlying benthos. The program can simulate time-varying processes of advection, dispersion, point and diffuse mass loading, boundary exchange, and kinetics of various water quality constituents. The model calculates variable concentrations for DO, biochemical oxygen demand (BOD), phytoplankton carbon and chlorophyll *a*, and nutrients. WASP5 must be coupled externally with a hydrodynamic model to simulate stratified lakes/reservoirs.

MIKE 3 is a fully three-dimensional hydrodynamic and water quality model. It comprises a series of modules that simulate hydrodynamics, advection and dispersion, water quality, eutrophication, and sediment transport. The variables modeled include BOD, ammonia, nitrate, DO, phosphorous, coliforms, phytoplankton, chlorophyll *a*, zooplankton, detritus, inorganic nitrogen, inorganic phosphorus, DO and benthic vegetation carbon, and bacteria.

CAEDYM is an ecological model developed by the Centre for Water Research (CWR) at the University of Western Australia. The model is designed to link to a number of hydrodynamic models, or it can be used alone if spatial variability can be ignored. The CAEDYM model was designed primarily to investigate eutrophication problems, and includes simulations of nutrient, phytoplankton, zooplankton, DO, and other state variables using temporal and spatial scales selected by the user. An external hydrodynamic model must be used to provide information on the flow field.

Three-dimensional dynamic process models can provide very detailed results for the spatial and time scales selected by the user. These models use detailed equations to simulate water quality kinetics and predict concentrations of important water quality parameters. However, more detailed or more precise results are not necessarily more accurate. Sufficient input and calibration data are needed to ensure meaningful results.

All four of the models reviewed are capable of simulating the processes and parameters important in reclaimed water reservoirs, including stratification, nutrient and DO cycles, algae growth, and sediment-water interactions. The models would be capable of simulating changes in nutrient inputs from reclaimed water, the watershed, and internal sources. In addition, they could be used to examine the effects of other management strategies, such as aeration systems to increase hypolimnetic DO, dredging, and alum treatment.

The three-dimensional models require significant technical expertise of the user to specify initial parameters, calibrate the model, and interpret the results. Technical expertise in the areas of modeling, limnology, and aquatic chemistry and aquatic biology would be required. Although the modeling capabilities of these models are significant, the data and user requirements would be too extensive for most utility managers with reclaimed water systems.

7.6 Model Selection

7.6.1 Model Evaluation Factors

Water quality models were evaluated in three areas:

- Ability to simulate conditions present in reclaimed water reservoirs;
- Effectiveness in addressing the issues of key water quality parameters, water quality criteria, and management strategies;
- Model requirements and limitations.

Because of the range of facilities, operations, and end uses of reclaimed water and because of problems encountered, there is no one perfect model for open reclaimed water reservoirs. However, models are available that can assist utility managers to better understand the water quality processes in their reservoirs and to evaluate the effects of management strategies on water quality in their reservoirs.

A water quality model can be used in a diagnostic or predictive mode. In a diagnostic mode, it is used to help understand the processes occurring in the reservoir. In the predictive mode, it is used to predict changes that would occur in the reservoir as a result of different conditions. The primary factors that should be considered in model selection include:

- Specific water quality problems encountered in the reservoir;
- Management strategies to be evaluated;
- Resources available for modeling including data that are currently available or that will be collected, technical expertise of the modeler, and time available for the modeling work.

The accuracy of any of the reservoir water quality models will depend to some extent on the amount of data available for model calibration and validation. Ideally, two or more data sets should be used in the calibration process to adjust rate coefficients and/or scaling factors. Model validation is another important step in determining the accuracy of the model. Model validation is accomplished by using the calibrated model to simulate other conditions represented by data that were not used in the calibration process. The accuracy of the model can be predicted by comparing the values estimated by the model to measured data.

7.6.2 Evaluation of Water Quality Problems and Management Strategies

As previously stated, the primary cause of water quality problems in reclaimed water reservoirs is excessive nutrient concentrations, and the two primary water quality problems are low DO and excessive algae. Management strategies are designed to either address the cause of the problems (i.e., the various sources of nutrients) or treat the water quality problems directly. The ability of the various water quality models to address these issues is discussed in this section.

Controlling the nutrient loading into the reservoir can be difficult and expensive. There are three sources of nutrients in open reclaimed water storage reservoirs: the reclaimed water, watershed runoff, and internal sources. The relative importance of each source will vary for each reservoir. The nutrient contribution from each source can be determined from monitoring data. A water quality model can be used to assess the relative importance of each source. However, if nutrient control is selected as the management strategy, all significant sources of nutrients should be controlled to ensure effectiveness. Management strategies to accomplish nutrient control for each of the three sources are listed in Table 7-1. A water quality model can be a useful tool to determine the extent of nutrient control that would be required to achieve the desired result in the reservoir.

With a water quality model, managers can estimate the extent to which each source of nutrients should be controlled. Most of the models reviewed are capable of assessing some aspects of nutrient reduction strategies. To evaluate nutrient management strategies, a model can be used to calculate in-reservoir nutrient concentrations resulting from nutrient loading from the significant sources.

Most of the models that predict in-reservoir nutrient concentrations will also predict chlorophyll *a* concentrations or Secchi disk depth. Some of the models, however, provide more detailed information on algae populations. For example, some of the models predict populations of various groups of algae and other biological groups; some predict the probability that blue-green algae will dominate.

Aeration either through destratification or oxygenation has been a very effective way to increase DO in reservoirs and alleviate hydrogen sulfide odors. In some cases, aeration has also reduced algae concentrations. To evaluate low DO problems, models should be able to simulate stratification and the processes that affect DO concentrations, such as BOD decay, sediment oxygen demand, and reaeration.

The ability of the various models to simulate water quality problems and management strategies is summarized in Table 7-2. The models are listed generally in order of complexity. The information in Table 7-2 can be used as a guide to determine which models could be used to address specific water quality problems. The models capable of simulating nutrient cycles and evaluating external nutrient reductions are indicated on the table (Y). The relative abilities of the models to address other problems and management strategies are indicated on the table using the terms basic (B), medium (M), and advanced (A). Although these relative rankings are subjective, they do provide a way to compare model capabilities. Each model has unique characteristics, as described in Chapter 6.0 and summarized on Table 6-1.

7.7 Summary

The typical water quality problems encountered in reclaimed water storage reservoirs are excess nutrients and algae, low DO, high turbidity, odors, and color. These problems do not occur independently, but are interrelated. Management strategies are designed to either address the cause of the problems, i.e., the various sources of nutrients, or treat the water quality problems directly. The management strategies that can be used to address excessive nutrients, low DO, and excessive algae are listed in Table 7-1.

Three of the models, EUTROMOD, PHOSMOD, and BATHTUB, are steady state empirical models. Five of the models, CE-QUAL-R1, CE-QUAL-W2, MIKE 11 RESERVOIR, MIKE 21, and RMA2 and 4, fit into the category of one-or two-dimensional dynamic process models. Four of the models, CE-QUAL-ICM, WASP5, MIKE 3, and CAEDYM, fit into the category of three-dimensional dynamic process models.

Because of the range of facilities, operations and end uses of reclaimed water, and the types of problems encountered, there is no one perfect model for reclaimed water storage reservoirs. With a water quality model, managers can estimate the extent to which each source of nutrients should be controlled. Most of the models reviewed can be used to examine some aspects of nutrient reduction strategies. The ability of the various models to simulate water quality problems and management strategies is summarized in Table 7-2.

Table 7-2. Water Quality Models Applicable to Water Quality Problems and Management Strategies

Model	Type	Water Quality Problems				Management Strategies				
		Nutrients	Low DO	Algae	Other	Reduce Nutrient Loads			Low DO	Algae
						Reclaimed Water	Watershed Runoff	Internal Sources		
EUTROMOD	well-mixed, steady state, empirical	Y		B		Y	Y			
PHOSMOD	well-mixed, steady state, empirical	Y	B			Y	Y	B		
BATHTUB	well-mixed segments, steady state, empirical	Y	B	B		Y	Y			
CE-QUAL-R1	1-D (vertical), dynamic, process	Y	M	M	Y	Y	Y		B	B
MIKE 21	2-D (horizontal), dynamic, process	Y		M	Y	Y	Y			M
RMA4	2-D (horizontal), dynamic, process				Y					
MIKE 11 RESERVOIR	2-D (vertical, longitudinal), dynamic, process	Y	A	M	Y	Y	Y	M	A	M
CE-QUAL-W2	2-D (vertical, longitudinal), dynamic, process	Y	A	M	Y	Y	Y	M	A	M
WASP5	3-D, external dynamic model, process	Y	A	M	Y	Y	Y	A	A	M
CE-QUAL-ICM	3-D, external dynamic model, process	Y	A	A	Y	Y	Y	A	A	A
MIKE 3	3-D, dynamic model, process	Y	A	M	Y	Y	Y	A	A	M
CAEDYM	3-D, external dynamic model, process	Y	A	A	Y	Y	Y	A	A	A

A = Advanced ability to address water quality problems and management strategies.
B = Basic ability to address water quality problems and management strategies.
M = Medium ability to address water quality problems and management strategies.
Y = Capable of simulating nutrient cycles and evaluating external nutrient reductions.

The study included 1) a review of existing information on management strategies that address water quality issues in open reclaimed water reservoirs, and 2) a review of existing models that could simulate water quality problems and the effectiveness of applying various management strategies to a reservoir with water quality problems. These reviews were completed in Chapters 5.0 and 6.0. The information was synthesized in Chapter 7.0 so that utility managers would be able to follow a logical path to determine which management strategy may successfully address their water quality problem(s) and which water quality model may be useful to understand specific water quality problems and simulate the effects of management strategies.

REFERENCES

<u>Note</u>: References with an asterisk (*) are described more fully in Appendix A.

*Allen and Kramer (1972). *Nutrients in Natural Waters.* John Wiley and Sons, New York, NY, pp. 14, 82.

*Ambrose, R.B., T.A. Wool, and J.L. Martin (1993). *The Water Quality Analysis Simulation Program, WASP5 Version 5.10. Part A: Model Documentation.* U.S. Environmental Protection Agency, Office of Research and Development, Environmental Research Laboratory, Athens, GA.

Ammerman, D.K. (1995). Evaluating Reuse Feasibility with Predictive Models. *Proc. 22nd Ann. Conf., Integrated Water Resour. Plan. for the 21st Century.* ASCE, Cambridge, MA, May 7-11, 1969.

*Asano, T., ed. (1998). *Water Quality Management Library—Vol. 10/Wastewater Reclamation and Reuse.* Technomic Publishing Co., Inc., Lancaster, PA.

*Ashby, S.L., and R.H. Kennedy (1999). *Water Quality Assessment for the Proposed Water Supply Reservoir, Duck River, Cullman, Alabama.* Technical Report EL-99-5. U.S. Army Engineer Waterways Experiment Station, Vicksburg, MS.

Barnard, James (2002). personal communication.

*Barnese, L.E., and J.A. Bohannon (1994). The Distribution of Nutrients and Phytoplankton in Taylorsville Lake—A Model Study. *In Symposium Proceedings on Responses to Changing Multiple-Use Demands: New Directions for Water Resources Planning and Management.* AWWA, Nashville, TN, April 17-20, 1994, pp. 33–35.

Barry, D., L.W. Melton, M. Ysusi, and K. Israel (1998). Initiation of Operations of California's Largest Recycled Water Program for Agriculture. *Proc. Joint AWWA/WEF Reuse Conference,* Lake Buena Vista, FL, 261.

*Baumgartner, D.J., W.E. Frick, and P.J.W. Roberts (1994). *Dilution Models for Effluent Discharges, 3rd ed.* EPA/600/R-93/139. U.S. Environmental Protection Agency, Newport, OR.

Benisch, M., R. Baur, M. Mengelkoch, W. Muirhead, J.B. Neethling, H. Stephens, B. Willey, and N. Cullen (2001). An Oregon Facility Tweaks Its Enhanced Biological Phosphorus Removal System to Cut Costs and Improve Compliance, *Water Environment & Technology* 13(4):38–44.

*Benndorf, J., and K. Putz (1987a). Control of Eutrophication of Lakes and Reservoirs by Means of Pre-Dam. I. Mode of Operation and Calculation of Nutrient Elimination Capacity. *Water Research* 21:829–38.

*Benndorf, J., and K. Putz (1987b). Control of Eutrophication of Lakes and Reservoirs by Means of Pre-Dams. II. Validation of the Phosphate Removal Model and Size Optimization. *Water Research* 21:839–42.

*Bernhardt, H. (1980). Reservoir Protection by In-River Nutrient Reduction. In: *Restoration of Lakes and Inland Waters.* EPA 440/5-81-010, pp. 272–77.

*Bernhardt, H., and J. Clasen (1985). Recent Developments and Perspectives of Restoration for Artificial Basins Used for Water Supply. *Conference Proceedings from Lake Pollution and Recovery:* European Water Pollution Control Association International Congress, Rome, Italy, pp. 213–27.

Berry, J.L., and S.A. Kacmar (1998). San Antonio Water System: Water Recycling Program. *Proc. Joint AWWA/WEF Water Reuse Conference,* Lake Buena Vista, FL, 533.

*Berzins, A., J. Lee, and E. Jackson (1998). Combined Water Supply and Wastewater Utilization. *Proc. WEF Annual Conference*, Orlando, FL.

*Bloetscher, F., P.A. Davis, C.W. Walker, and A. Muniz (1998). Innovative Uses of Reclaimed Water Reduce Stress on Hollywood Water Supply. *Proc. AWWA Annual Conference*, Dallas, TX, B, 489.

*BOSS International (2001a). SMS RMA2. http://www.bossintl.com/html/rma2.html

*BOSS International (2001b). SMS RMA4. http://www.bossintl.com/html/rma4.html

*Bowie, G.L., W.B. Mills, D.B. Porcella, C.L. Campbell, J.R. Pagenkopf, G.L. Rupp, K.M. Johnson, P.W.H. Chan, and S.A. Gherini (1985). *Rates, Constants, and Kinetic Formulations in Surface Water Quality Modeling, 2nd ed.* EPA/600/3-85/040 U.S. Environmental Protection Agency, Environmental Research Laboratory, Athens, GA.

*Brooks, J.L., and S.I. Dodson (1965). Predation, Body Size, and Composition of Plankton. *Science* 150:28–35.

*California Potable Reuse Committee (1996). *A Proposed Framework for Regulating Indirect Potable Reuse of Advanced Treated Reclaimed Water by Surface Water Augmentation in California*. California Department of Water Resources, Sacramento, CA.

*Camp, Dresser & McKee, Inc. (1983). *Cross Lake Nutrient Removal Study, Preliminary Dredging Assessment*. Submitted to City of Shreveport, LA.

*Casitas Municipal Water District (1987). *Current Methodology for the Control of Algae in Surface Reservoirs.* AWWA Research Foundation, Denver, CO.

*Cerco, C.F., and T. Cole (1993). Three-dimensional Eutrophication Model of the Chesapeake Bay. *Journal of Environmental Engineering* 119(6):1006–25.

*Cerco, C.F., and T. Cole (1995). *User's Guide to the CE-QUAL-ICM. Release Version 1.0.* Technical Report. EL-95-15. U.S. Army Engineer Waterways Experiment Station, Vicksburg, MS.

*Chapra, S., and R.P. Canale (1991). Long-term Phenomenological Model of Phosphorus and Oxygen for Stratified Lakes. *Water Research* 25(6):707–15.

*Cheng, C., J.F. Atkinson, and J.V. DePinto (1994). A Coupled GIS-Water Quality Modeling Study. In: *Proceedings of the 1994 Hydraulic Engineering Conference*, American Society of Civil Engineers, Buffalo, New York, NY, pp. 247–51.

*Chu, S. P. (1943). The Influence of the Mineral Composition of the Medium on the Growth of Planktonic Algae. II. The Influence of the Concentration of Inorganic Nitrogen and Phosphate Phosphorous. *J. Ecol.* 31:109–48.

*Cole, R.W., and E.M. Buchak (1995). *CE-QUAL-W2: A Two-dimensional, Laterally Average, Hydrodynamic and Water Quality Model.* Version 2.0 Instructional Report EL-95-1. U.S. Army Engineer Waterways Experiment Station, Vicksburg, MS.

*Cole, T.M., and D.H. Tillman (1997). *Water Quality Modeling of Lake Monroe Using CE-QUAL-W2*, Miscellaneous Paper EL-99-1. U.S. Army Engineer Waterways Experiment Station, Vicksburg, MS.

*Cooke, G.D., and R.E. Carlson (1986). Water Quality Management in a Drinking Water Reservoir. *Lake Reserv. Management* 2:363–71.

*Cooke, G.D., and R.E. Carlson (1989). *Reservoir Management for Water Quality and THM Precursor Control.* AWWA Research Foundation, Denver, CO, 387 pp.

*Cooke, G.D., E.B. Welch, S.A. Peterson, and P.R. Newroth (1993). *Restoration and Management of Lakes and Reservoirs, Second Edition.* Lewis Publishers, Boca Raton, FL.

*Crook, J., G.F. Shimp, and J. Sandino (1996). Water Reuse Strategy for Santiago, Chile. *Proc. Joint AWWA/WEF Water Reuse Conference*, San Diego, CA.

Crook, J. (1998). Water Reclamation and Reuse Criteria. In: *Wastewater Reclamation and Reuse*, pp. 627–703, T. Asano (ed.). Technomic Publishing Co., Ltd., Lancaster, PA.

Crook, J., L.J. Johnson, and K. Thompson (2001). California's New Water Recycling Criteria and Their Effect on Operating Agencies. In: *Proceedings of The American Water Works Association 2002 Annual Conference*, June 17–21, 2001, Washington, D.C.

*Cuff, Denis (1993). Oxygen Pump Aims to Save Reservoir Fish in Stale Water. *Contra Costa Times* June 22.

*Cullen, P., and C. Forsberg (1988). Experiences with Reducing Point Sources of Phosphorus to Lakes, *Hydrobiologia* 170:321–36.

*Danish Hydraulic Institute (2000a). *MIKE 11 Short Description*. Danish Hydraulic Institute, Horsholm, Denmark.

*Danish Hydraulic Institute (2000b). *MIKE 21 Environmental Modelling—A Short Description.* Danish Hydraulic Institute, Horsholm, Denmark.

*Danish Hydraulic Institute (2000c). MIKE 3, *A Three-dimensional Hydrodynamic Model, A Short Description*. Danish Hydraulic Institute, Horsholm, Denmark.

*Delfino, J.J., and G.F. Lee (1971). Variation of Manganese, Dissolved Oxygen and Related Chemical Parameters in the Bottom Waters of Lake Mendota, Wisconsin. *Water Research* 5:1207.

*Dunst, R.C. (1981). Dredging Activities in Wisconsin's Lake Renewal Program. In: *Restoration of Lake and Inland Waters: International Symposium on Inland Waters and Lake Restoration*, EPA 440/5-81-010.

*Dunst, R.C., J.G. Vennie, R.B. Corey, and A.E. Peterson (1984). Effect of Dredging Lilly Lake, Wisconsin. EPA 600/3-84-097.

*Ernst, M.R., W. Frossard and J.L. Mancini (1994). *Two Eutrophication Models Make the Grade*, *Water Environment & Technology* November, 1994 pp. 15–16.

*Fast, A.W. (1979). Artificial Aeration as a Lake Restoration Technique. In: *Lake Restoration.* EPA 440/5-79-001, pp. 121–132.

*Ferrara, T.A., and A. Hildick-Smith (1982). A Modeling Approach for Storm Water Quantity and Quality Control Via Detention Basins. *Water Resources Bulletin* 18:975–81.

*Florida Department of Environmental Protection (1996). *Reuse of Reclaimed Water and Land Application*. Chapter 62-610, Florida Administrative Code. FDEP, Tallahassee, FL.

*Gambrell, R.P., C.N. Reddy, and R.A. Khalid (1983). Characterization of Trace and Toxic Materials in Sediments of a Lake Being Restored. *J. Water Poll. Cont. Fed.* 55:1201–10.

Goldman, C.R., and A.J. Horne (1983). *Limnology*, McGraw-Hill, New York, NY.

*Harvey MUDD College Engineering Clinic (1993). *Final Report to Irvine Ranch Water District, Reclaimed and Nonpotable Water System Analysis and Recommended Strategies to Water Quality Problems.* Claremont, CA. Engineering Clinic.

*Hayes, C.R., R.G. Clark, R.F. Stent, and C.J. Redshaw (1984). The Control of Algae by Chemical Treatment in a Eutrophic Water Supply Reservoir. *J. Inst. Water Eng. Sci.*, 38:149–62.

*Herbel, C.H. (1982). Grazing Management on Rangelands, *Journal of Soil and Water Conservation* 37(2):77–79.

*Hernandez, P., R.B. Ambrose Jr., D. Prats, E. Ferrandis, and J.C. Asensi (1997). Modeling Eutrophication Kinetics in Reservoir Microcosms. *Water Research* 31(10):2511–19.

*Herzfeld, M., and D. Hamilton (2000a). *The CWR Computational Aquatic Ecosystem Dynamics Model, CAEDYM, Science Manual*. Center for Water Research, The University of Western Australia, September 25. http://www.cwr.uwa.edu.au/~antenucc/cd/cwrsoft/doc/caedym_user/title.html

*Herzfeld, M., and D. Hamilton (2000b). *The CWR Computational Aquatic Ecosystem Dynamics Model, CAEDYM, User's Manual*. Center for Water Research, The University of Western Australia, September 25. http://www.cwr.uwa.edu.au/~antenucc/cd/cwrsoft/doc/caedym_user/title.html

*Hutchinson, G. E. (1957). *A Treatise on Limnology*. Vol. 1. Geography, Physics and Chemistry. John Wiley and Sons, New York, NY.

*Hutchinson, G. E., and H. Lofler (1956). The Thermal Classification of Lakes. *Proceedings National Academy of Science* 42.

*Irvine Ranch Water District and Black & Veatch (1994). *Final Report Watershed Management Study for Rattlesnake and Sand Canyon Reservoirs*, Irvine, CA.

Jordan, B. (2001). *Water Quality Data from Santa Margarita Water District for Upper Oso Reservoir* (oral communication).

*Kennedy, R.H. (1995). *Application of the BATHTUB Model to Selected Southeastern Reservoirs.* Technical Report EL-95-14. U.S. Army Engineer Waterways Experiment Station, Vicksburg, MS.

*Landers, J.C., and B.A. Knuth (1991). Use of Wetlands for Water Quality Improvement Under the U.S. EPA Region V Clean Lakes Program. *Environ. Manage.* 15:151–62.

*Lang, G.A., and T.D. Fontaine (1990). Modeling the Fate and Transport of Organic Contaminants in Lake St. Clair. *Journal of Great Lakes Research* 16(2):216–32.

*Lowe, E.F., L.E. Battoe, D.O. Stites, and M.F. Coveney (1992). Particulate Phosphorus Removal Via Wetland Filtration: an Examination of Potential for Hypertrophic Lake Restoration. *Environ. Manage.* 16:67–74.

*Lung, W., and C.E. Larson (1995). Water Quality Modeling of the Upper Mississippi River and Lake Pepin. *Journal of Environmental Engineering* 121 (10):691–99.

*Lusby, G.C. (1989). *Effects of Grazing on Runoff and Sediment Yield from Dessert Rangeland at Badger Wash in Western Colorado, 1953–1973.* USES Water Supply Paper 1532-1, Washington, DC.

*Lynch, S., K. Martin, and D. Bramwell (1998). Southern California Comprehensive Water Reclamation and Reuse Study. *Proc. WEF Watershed Specialty Conference*, Denver, CO, 1227.

*Martin, E.H. (1988). Effectiveness of an Urban Runoff Detention Pond-Wetland System. *Journal of Environmental Engineering* 114 (3):810–27.

*McGinn, J., L. Barbato, J. Van Teylingen, and V. L. Card (1998). Colorado Springs: A Model for Reclaimed Water Systems. *Proc. Joint AWWA/WEF Water Reuse Conference*, Lake Buena Vista, FL, 497.

*McKnight, D.M., S.W. Chisholm, and D.R.F. Harleman (1983). CuSO4 Treatment of Nuisance Algal Blooms in Drinking Water Reservoirs. *Environ. Manage.* 7:311–20.

*Meyer, B. (1991). Pneumatic Destratification System Design Using a Spreadsheet Program. *Information Exchange Bulletin*, Vol. E-91-1, WOTS U.S. Army Corps Engineers, Waterways Experiment Station, Vicksburg, MS.

Mitchell, R. (1972). *Water Pollution Microbiology*. Wiley Interscience, pp. 32, 38.

*Mortimer, C.H. (1941). The Exchange of Dissolved Substances between Mud and Water in Lakes. 1 and 2. *J. Ecol.* 29:280–329.

* Mortimer, C.H. (1942). The Exchange of Dissolved Substances between Mud and Water in Lakes. 3 and 4. *J. Ecol.* 30:147–201.

* Mortimer, C.H. (1971). Chemical Exchanges Between Sediments and Water in the Great Lakes-Speculations on Probable Regulatory Mechanisms. *Limnol. Oceanography* 16:387–404.

*National Research Council (1998). *Issues in Potable Reuse*. National Academy Press, Washington, DC.

*Nichols, D.S. (1983). Capacity of Natural Wetlands to Remove Nutrients from Wastewater. *Journal WPCF* 55:495–505.

*Noble, R.L. (1986). Predator-Prey Interactions in Reservoir Communities. In: *Reservoir Fisheries Management: Strategies for the 80's*. G.E. Hall and M.J. Van Den Avyle (eds.), Reservoir Comm. SDAFS, Bethesda, MD.

*Oberts, G.L., and R.A. Osgood (1991). Water-quality Effectiveness of a Detention/Wetland Treatment System and Its Effect on an Urban Lake. *Environ. Manage.* 15:131–38.

*Olem, H., and G. Flock, eds. (1990). *Lake and Reservoir Restoration Guidance Manual, 2nd edition*. EPA 440/4-90-006. Prepared by North American Lake Management Society for U.S. Environmental Protection Agency, Washington, D.C.

*Owens, L.B., W.M. Edwards, and R.W. Van Keuren (1989). Sediment and Nutrient Losses from an Unimproved All-Year Grazed Watershed. *J. Environ. Qual.* 18:232.

*Palmer, M. (2001). *Water Quality Modeling*. The World Bank, Washington, D.C.

*Parker, Sybil P. (1984). *Concise Encyclopedia of Science and Technology*. McGraw-Hill Book Company, New York, NY, p. 354.

*Pastorak, R.A., M.W. Lorenzen, and T.C. Ginn (1982). *Environmental Aspects of Artificial Aeration/and Oxygenation of Reservoirs: A Review of Theory, Techniques, and Experiences*. Technical Report No. E-82-3. U.S. Army Corps Engineers Waterways Exp. Sta., Vicksburg, MS.

*Peterson, S.A. (1982). Lake Restoration by Sediment Removal. *Water Resources Bulletin* 18:423–35.

*Prato, T., and H. Shi (1990). A Comparison of Erosion and Water Pollution Control Strategies for an Agricultural Watershed. *Water Resources Research* 26(2):199–205.

*Raman, R.K., and R.L. Evans (1984). *Aeration-Destratification of Lake Eureka: Second Year Operation*. ISWS/CIR-159/84, Circular 159, State of Illinois, Department of Energy and Natural Resources.

*Rausch, D.L., and J.D. Schreibe (1981). Sediment and Nutrient Trap Efficiency of a Small Flood-Detention Reservoir. *Journal of Environmental Quality* 10(3):288–93.

*Reckhow, K.H. (1990). *EUTROMOD Spreadsheet Program—A Regional Modeling Scheme for Nutrient Runoff and Lake Trophic State Modeling*. School of Forestry and Environmental Studies, Duke University, Durham, NC.

*Robbins, R.W., J.L. Glicker, D.M. Bloem, and B.M. Niss (1991). *Effective Watershed Management for Surface Water Supplies*. AWWA Research Foundation, Denver, CO, 401 pp.

*Rodhe, W. (1948). Environmental Requirements of Freshwater Plankton Algae. Experimental Studies in the Ecology of Phytoplankton. *Symbol. Bot. Upsalien* 10(1), 149 pp.

Rowe, D.R., and I.M. Abdel-Magid (1995). *Handbook of Wastewater Reclamation and Reuse.* Lewis Publishers, Boca Raton, FL.

*Sartoris, J.J., and J.T. Boehmke (1987). *Limnological Effects of Artificial Aeration at Lake Cachuma, California, 1980–1984.* REC-ERC-87-10. U.S. Department of the Interior, Bureau of Reclamation.

Sas, H., I. Ahlgren, H. Bernhardt, B. Bostrom, J. Clasen, C. Forsberg, D. Imboden, L. Kamp-Nielson, L. Mur, N. de Oude, C. Reynolds, H. Schreurs, U. Sommer, and S. Vermij (1989). *Lake Restoration by Reduction of Nutrient Loading: Expectations, Experiences, Extrapolation.* Academia-Verlag, Richarz, St. Augustine, Germany.

*Schladow, S.G. (1993). Lake Destratification by Bubble-Plume Systems: Design Methodology. *Journal of Hydraulic Engineering* 119 (3):350–68.

*Schueler, T.R. (1987). *Controlling Urban Runoff: A Practical Manual for Planning and Designing Urban BMPs*. Washington, D.C., Metropolitan Washington Council of Governments.

*Shoemaker, L., M. Lahlou, M. Bryer, D. Kumar, and K. Kratt (1997). *Compendium of Tools for Watershed Assessment and TMDL Development.* EPA841-B-97-006. U.S. Environmental Protection Agency, Office of Water, Washington, D.C.

Smith, V. H. (1979). Nutrient Dependence of Primary Productivity in Lakes. *Limnol. Oceanography* 24:1051–64.

Spangenberg, C., and A. Kalinsky (1993). *MWRP Secondary Effluent Phosphorous Removal Microfiltration Study*. Irvine Ranch Water District.

*Spangenberg, C. (1991). *Assessment of Hypolinmetic Aeration as a Restorative Technique for Irvine Lake*. Thesis for California State University, Long Beach.

*Speece, Richard E. (1994). Lateral Thinking Solves Stratification Problems. *WQI* 3:12–15.

*Speece, R.E. (1996). Oxygen Supplementation by U-Tube to the Tombigbee River. *Wat. Sci. Tech.* 34(12):83–90.

*Speece, R.E. (1989). *Tailwaters: A Regulated Streams' Most Valuable Resource*. Prepared for North American Lake Management Society Plenary Session.

*Speece, R.E. (2000). *Proposed Design of an Oxygen Supplementation System for Percy Priest Reservoir*. Prepared for the Nashville U.S. Army Corp of Engineers.

*Speece, R.E., N. Nirmalakhandan, and G. Tchobanoglous (1990). Commercial Oxygen Use in Water-Quality Management. *Water Environment & Technology*, July:54–61.

*Striegl, R.G. (1987). Suspended Sediment and Metals Removal from Urban Runoff by a Small Lake. *Water Resources Bulletin* 23(6):985–96.

*Stumm, W., and J.J. Morgan (1970). *Aquatic Chemistry*. Wiley Interscience, New York, NY.

*Tchobanogolous, G. (1985). *Water Quality*. Addison-Wesley, USA, pp. 121–23, 188–89.

Tchobanogolous, G., F.L. Burton, and H.D. Stensel (2003). *Wastewater Engineering: Treatment and Reuse*, McGraw-Hill, New York, NY.

*Tetra Tech, Inc. (1995). *Watershed Protection: Clean Lakes Case Study, Phosphorus Inactivation and Wetland Manipulation, Kezar Lake, NH*. EPA 841-F-95-002.

*Thomann, R.V., and J.A. Mueller (1987). *Principles of Surface Water Quality Modeling and Control*. Harper & Row, New York, NY.

*Trott, K.E., and M.J. Singer (1983). Relative Erodibility of 20 California Range and Forest Soils. *Journal of Soil Science of America* 47:753–59.

*Tufford, D.L., and H.N. McKeller (1999). Spatial and Temporal Hydrodynamic and Water Quality Modeling Analysis of a Large Reservoir on the South Carolina Coastal Plain. *Ecological Modeling* 114.

*Uhlmann, D. and H. Klapper (1985). Protection and Restoration of Lakes and Reservoirs in the German Democratic Republic—Case Studies. *Conference proceedings from Lake Pollution and Recovery:* European Water Pollution Control Association International Congress, Rome, Italy, pp. 262–71.

*U.S. Army Corps of Engineers (1982). *CE-QUAL-R1: A Numerical One-Dimensional Model of Reservoir Water Quality; User's Manual*. Instruction Report E-82-1 revised edition. U.S. Army Engineer Waterways Experiment Station Environmental Laboratory, Vicksburg, MS, 623 pp.

*U.S. Army Corps of Engineers (2000). *CE-QUAL-R1, One-Dimensional (Vertical) Reservoir Water Quality Model*. Environmental Laboratory, Vicksburg, MS. http://www.wes.army.mil/el.elmodels/r1info.html.

*U.S. EPA (1984). *Technical Guidance Manual for Performing Waste Load Allocations—Book IV Lakes and Impoundments—Chapter 2 Nurtient/Eutrophication Impacts*. EPA 440/4-84-019. U.S. Environmental Protection Agency, Office of Water Regulations and Standards. Washington, D.C.

*U.S. EPA (1986). *Quality Criteria for Water, 1986.* PB87-226759. U.S. Environmental Protection Agency, Washington, D.C.

*U.S. EPA (1988). *Design Manual, Constructed Wetlands and Aquatic Plant Systems for Municipal Wastewater Treatmen., EPA/625/1-88/022.* U.S. Environmental Protection Agency, Washington, D.C.

*U.S. EPA (1991). *Municipal Water Reuse, Selected Readings on Water Reuse.* U.S. EPA 430/9-91-022. U.S. Environmental Protection Agency, Washington, D.C.

*U.S. EPA (1992). *Guidelines for Water Reuse*. EPA/625/R-92/004. U.S. EPA, Center for Environmental Research Information, Cincinnati, OH.

*U.S. EPA (1995). *Watershed Protection: Clean Lakes Case Study, Watershed and In-Lake Practices Improve Green Valley Lake, Iowa*. EPA 841-F-95-003.

*Walker, W.W. (1987). *Empirical Methods for Predicting Eutrophication in Impoundments.* Report 4, Phase III, Application Manual, Technical Report E-81-9. U.S. Army Corps of Engineers, Vicksburg, MS.

*Walker, W.W. (1996). *Simplified Procedures for Eutrophication Assessment and Prediction: User Manua.,* Instruction Report W-96-2. U.S. Army Corps of Engineers, Waterways Experiment Station, Water Operations Technical Support Program.

*Washington State Department of Health (1997). *Water Reclamation and Reuse Standards.* WDOH, Olympia, WA.

*Wedepohl, R.E., D.R. Knauer, G.B. Wolbert, H. Olem, P.J. Garrison, and K. Kepford (1990). *Monitoring Lake and Reservoir Restoration*. EPA 440/4-90-007. Prepared by N. Am. Lake Manage. Soc. for U.S. Environmental Protection Agency, Office of Water, Washington, D.C.

Welch, M.R., I. Hannoun, R. Triessell, P. Gagliardo, and K. Weinberg (1996). *Storage of Repurified Water in a Domestic Supply Reservoir.* Paper presented at Joint AWWA/WEF Water Reuse Conf., San Diego, CA.

*Wen, J., and R.S. Torrest (1987). Aeration-Induced Circulation from Line Sources. I: Channel Flows. *J. of Environmental Engineering* 113 (1):82–98.

*Wen, J., and R.S. Torrest (1987). Aeration-Induced Circulation from Line Sources. II: Dissolved Oxygen Variations. *J. of Environmental Engineering* 113 (1):99–105.

*Wetzel, R. (1975). *Limnology*. W. B. Saunders Company, Philadelphia, PA, pp. 78–79, 215–45.

*Wetzel, R. (1983). *Limnology*. 2nd ed. Saunders. Philadelphia, PA.

APPENDIX A

Literature Review

Various search tools were accessed using key words applicable to water quality models, water quality criteria, and reservoir management techniques. The search tools included COMPENDEX, Applied Science and Technology, Environment, and WORLDCAT. In addition, several web sites were searched, including the Environmental Protection Agency (www.epa.gov), the American Water Works Association (www.awwa.org), the American Society of Civil Engineers (www.asce.org), the Water Environment Federation (www.wef.org), and the United States Geological Survey (www.usgs.gov). The annual literature reviews published in the June issues of Water Environment and Technology for the years 1966 through 2000 were reviewed, and additional references were obtained from these literature reviews. The following is a bibliography of references considered useful to the study or to municipalities interested in the storage of reclaimed water.

Allen and Kramer (1972). *Nutrients in Natural Waters*. John Wiley and Sons, New York, pp. 14, 82.

> Critical nutrients that promote biological growth are identified and described. Sources of nutrients, transformation in water, sediment, and production of biomass are discussed.

Ambrose, R.B., T.A. Wool, and J.L. Martin (1993). *The Water Quality Analysis Simulation Program, WASP5 Version 5.10. Part A: Model Documentation.* U.S. Environmental Protection Agency, Office of Research and Development, Environmental Research Laboratory, Athens, GA.

> Documentation for the WASP5 model is provided including a description of the model components. The processes simulated by the model and the modeling parameters are discussed. Major components include chemical tracers, sediment transport, eutrophication, simple toxicants, and organic chemicals.

Asano, T., ed. (1998). *Water Quality Management Library—Vol. 10/Wastewater Reclamation and Reuse.* Technomic Publishing Co., Inc., Lancaster, PA.

Technical information on all aspects of water reuse and reclamation based on actual operating facilities across the world is provided. Includes design, implementation, operational, and water quality information on reuse and reclamation. Physical, chemical, and biological data requirements for understanding water reuse and its effects are included.

Ashby, S.L., and R.H. Kennedy (1999). *Water Quality Assessment for the Proposed Water Supply Reservoir, Duck River, Cullman, Alabama*, Technical Report EL-99-5, U.S. Army Engineer Waterways Experiment Station, Vicksburg, MS.

To assess the potential water quality problems in a proposed reservoir on the Duck River, a study was conducted using the BATHTUB model. The model was used to estimate reservoir water quality based on existing nutrient loads and reduced loads resulting from implementation of watershed Best Management Practices (BMPs). Results of the study are presented in this technical report.

Barnese, L.E., and J.A. Bohannon (1994). The Distribution of Nutrients and Phytoplankton in Taylorsville Lake—A Model Study. In: *Symposium Proceedings on Responses to Changing Multiple-Use Demands: New Directions for Water Resources Planning and Management.* AWWA, Nashville, TN, April 17-20, 1994, pp. 33–35.

Application of the CE-QUAL-W2 model to Taylorsville Lake in Kentucky is described. CE-QUAL-W2 is a two dimensional laterally average model.

Baumgartner, D.J., W.E. Frick, and P.J.W. Roberts (1994). *Dilution Models for Effluent Discharges, 3rd ed.* EPA/600/R-93/139. U.S. Environmental Protection Agency, Newport, OR.

Information for three dilution models, PLUMES, CORMIX and RSB (Roberts, Snyder and Baumgartner) is provided. In addition to specific information for each of the three models, a general discussion of dilution modeling is included. The report also provides recommendation on model selection.

Benndorf, J., and K. Putz (1987a). Control of Eutrophication of Lakes and Reservoirs by Means of Pre-dam. I. Mode of Operation and Calculation of Nutrient Elimination Capacity. *Water Research* 21:829–38.

A model to calculate the removal of orthophosphorus for pre-dams of a particular size is presented. The dominant process of nutrient removal in pre-dams is biological uptake and sedimentation. The paper also discusses how to estimate removal of inorganic nitrogen.

Benndorf, J., and K. Putz (1987b). Control of Eutrophication of Lakes and Reservoirs by Means of Pre-dams. II. Validation of the Phosphate Removal Model and Size Optimization. *Water Research* 21:839–42.

Companion paper to the one listed above, in which the validity of the model is checked using 17 data sets from 15 existing pre-dams. The model can also be used to determine the optimum size of a pre-dam for the required rate of nutrient removal. Optimum size pre-dams result in about a 60 percent removal of phosphorus.

Bernhardt, H. (1980). *Reservoir Protection by In-river Nutrient Reduction*, in *Restoration of Lakes and Inland Waters.* EPA 440/5-81-010, pp. 272–277.

Results of a study on the operation of a phosphorus elimination plant on the Wahnbach Reservoir are presented. Over a 90 percent removal rate was achieved for phosphorus, algae, turbidity, and bacteria. The trophic state of the reservoir shifted from eutrophic to oligitrophic.

Bernhardt, H., and J. Clasen (1985). *Recent Developments and Perspectives of Restoration for Artificial Basins Used for Water Supply.* Conference proceedings form Lake Pollution and Recovery: European Water Pollution Control Association, International Congress, Rome, Italy. pp. 213–27.

In-reservoir technical treatment techniques utilized in various basins located in Europe along with performance are described.

Berzins, A., J. Lee, and E. Jackson (1998). Combined Water Supply and Wastewater Utilization. *Proc. WEF Annual Conference*, Orlando, FL.

The arid areas of the Okanagan Valley of British Columbia required innovative solutions to the increasing urbanization pressures on the water supply and wastewater treatment. A dual water distribution system, drawing on the existing high quality Kalamalka Lake water for domestic demands in a future distribution network, and using highly treated wastewater, supplemented with lower quality creek waters, for urban irrigation and fire protection, would cost $35 M. Releasing effluent to Okanagan Lake and upgrading water supplies to drinking water standards separately would cost $85 M.

Bloetscher, F., P.A. Davis, C.W. Walker, and A. Muniz (1998). Innovative Uses of Reclaimed Water Reduce Stress on Hollywood Water Supply. *Proc. AWWA Annual Conference*, Dallas, TX, B, 489.

Hollywood, FL incorporated the needs of both the water and wastewater systems in an integrated water resource program. The direction that the city pursued is intended to reduce competition for the existing water supplies and to enable the city to maintain its current water supply allocations. Establishing a reuse system has reduced competition with the city's wellfield. The city now sees demand for reclaimed water from nontraditional sources, such as power plants and airports.

BOSS International (2001a). SMS RMA2. http://www.bossintl.com/html/rma2.html

RMA2 is a two-dimensional, depth-averaged finite element hydrodynamic numerical model that can be linked to the water quality model RMA4. The capabilities, limitation and simulation types of the model are described. It also provides other important model information such as boundary conditions, time steps and modeling storm and tidal conditions.

BOSS International (2001b). SMS RMA4. http://www.bossintl.com/html/rma4.html

RMA4 is a two-dimensional, depth-averaged finite element water quality transport numerical model that can simulate up to six constituents. The capabilities, limitation and simulation types of the model are described. It also provides other important model information such as specifying diffusion coefficients, assigning initial conditions, and specifying boundary conditions.

Bowie, G.L., W.B. Mills, D.B. Porcella, C.L. Campbell, J.R. Pagenkopf, G.L. Rupp, K.M. Johnson, P.W.H. Chan, and S.A. Gherini (1985). *Rates, Constants, and Kinetic Formulations in Surface Water Quality Modeling, 2nd ed.* EPA/600/3-85/040 U.S. Environmental Protection Agency, Environmental Research Laboratory, Athens, GA.

Formulations used in surface water modeling and accepted rate constants are provided in this report. Dispersion processes, the dissolved oxygen balance and the relationships between nutrients and algal growth are also discussed.

Brooks, J.L., and S.I. Dodson (1965). Predation, Body Size, and Composition of Plankton. *Science* 150:28–35.

The phytoplankton, zooplankton, and fish populations in several lakes in Connecticut are discussed. The absence of large Daphnia was observed in several lakes with natural alewives populations. Results showed that planktivorious fish grazed selectively on large bodied zooplankton. In lakes with large populations of planktivors, there were virtually no large bodied zooplankton and a larger biomass of phytoplankton.

California Potable Reuse Committee (1996). *A Proposed Framework for Regulating Indirect Potable Reuse of Advanced Treated Reclaimed Water by Surface Water Augmentation in California*. California Department of Water Resources, Sacramento, CA.

In June 2001, the State DOHS drafted revisions to Title 22 Standards for Water Recycling which shall be entitled, "The Purple Book: Excerpts from the Health and Safety Code, Water Code, and Titles 22 and 17 of the California Code of Regulations." The new regulations have incorporated a number of changes from the 1978 version. Requirements for treatment have changed significantly during the past twenty-eight years.

Camp, Dresser & McKee, Inc. (1983). *Cross Lake Nutrient Removal Study, Preliminary Dredging Assessment*, submitted to City of Shreveport, LA.

An assessment of the feasibility of removing bottom sediments from Cross Lake to eliminate or reduce the potential for nutrient releases into the water is presented. The work included a nutrient budget evaluation, data collection and analysis, and dredging assessment.

Casitas Municipal Water District (1987). *Current Methodology for the Control of Algae in Surface Reservoirs.* AWWA Research Foundation, Denver, CO.

The results of a study funded by the AWWA Research Foundation on the control of algae in surface water reservoirs are presented. Also included are a literature review of several algae control techniques, results of surveys of water utilities and chemical manufacturers, a discussion on copper speciation and chemistry, and results of field testing of copper compounds in Lake Casitas, California.

Cerco, C.F., and T. Cole (1993). Three-dimensional Eutrophication Model of the Chesapeake Bay. *Journal of Environmental Engineering* 119(6):1006–25.

A model package of the Chesapeake Bay that consists of three separate models coupled and integrated to predict the transport (CH3D-WES), water quality (CE-QUAL-ICM), and sediment processes of the system is presented and discussed. The complete model is a three-dimensional, time-variable, eutrophication model incorporating 22 state variables that describe the physical, biological, and chemical properties of the water body. The capability of modeling complex eutrophic systems with coupled three-dimensional hydrodynamic and water quality models is demonstrated in this study.

Cerco, C.F., and T. Cole (1995). *User's Guide to the CE-QUAL-ICM. Release Version 1.0.* Technical Report. EL-95-15. U.S. Army Engineer Waterways Experiment Station, Vicksburg, MS.

CE-QUAL-ICM is a three-dimensional eutrophication model. This document is the user's guide for the model and provides information on model development, equations used in the model, eutrophication kinetics, sediment-water interactions, and the mechanics of operating the model.

Chapra, S., and R.P. Canale (1991). Long-term Phenomenological Model of Phosphorus and Oxygen for Stratified Lakes. *Water Research* 25(6):707–15.

A model developed to predict the long-term response of a lake to changes in phosphorus loading is described. The total phosphorus budget is determined by external loading and a sediment recycle rate based upon the areal hypolimnetic oxygen demand (AHOD). The model is applied to Shagawa Lake, MN, for illustration with successful results. The authors conclude that the model can provide a framework for assessing the impact of sediment feedback on long-term lake recovery.

Cheng, C., J.F. Atkinson, and J.V. DePinto (1994). A Coupled GIS-Water Quality Modeling Study. In: *Proceedings of the 1994 Hydraulic Engineering Conference*, American Society of Civil Engineers, Buffalo, New York, pp. 247–51.

Lake restoration work conducted at Kezar Lake is summarized. Sediment phosphorus inactivation and wetlands management were implemented to address water quality problems. Results of post-restoration monitoring conducted over a 10-year period are summarized in the case study report.

Chu, S. P. (1943). The Influence of the Mineral Composition of the Medium on the Growth of Planktonic Algae. II. The Influence of the Concentration of Inorganic Nitrogen and Phosphate Phosphorous. *J. Ecol.* 31:109–48.

An assessment of the impact of nitrogen and phosphorous levels on the growth of planktonic algae and optimum levels for growth is presented.

Cole, R.W., and E.M. Buchak (1995). *CE-QUAL-W2: A Two-dimensional, Laterally Average, Hydrodynamic and Water Quality Model.* Version 2.0 Instructional Report EL-95-1, U.S. Army Engineer Waterways Experiment Station, Vicksburg, MS.

CE-QUAL-W2 is a two-dimensional lake water quality model. This document is the user's guide for the model and provides information on model development, model processes, equations used in the model, and the mechanics of operating the model.

Cole, T.M., and D.H. Tillman (1997). *Water Quality Modeling of Lake Monroe Using CE-QUAL-W2.* Miscellaneous Paper EL-99-1, U.S. Army Engineer Waterways Experiment Station, Vicksburg, MS.

The CE-QUAL-W2 model was used to investigate temperature and water quality for Lake Monroe, IN. The Louisville District of the Corps of Engineers needed a model to investigate future water quality conditions resulting from potential changes in reservoir operations and/ or changes in environmental conditions.

Cooke, G.D., and R.E. Carlson (1986). Water Quality Management in a Drinking Water Reservoir. *Lake Reserv. Management* 2:363–71.

An example of an in-stream treatment for Lake Rockwell is discussed. Aluminum sulfate was added to the river, which provides 90% of the water to the lake. Although treatment removed over half of the soluble reactive phosphate, total phosphorus loading to the lake was not reduced. In addition, aluminum floc built up rapidly on the river bed.

Cooke, G.D., and R.E. Carlson (1989). *Reservoir Management for Water Quality and THM Precursor Control.* AWWA Research Foundation, Denver, CO, 387pp.

This report focuses on drinking water reservoirs and contains information on water quality processes and on many lake restoration techniques. Techniques discussed in the report include dredging, destratification, copper sulfate, oxygenation, and biomanipulation. There are some very limited discussions of case studies.

Cooke, G.D., E.B. Welch, S.A. Peterson, and P.R. Newroth (1993). *Restoration and Management of Lakes and Reservoirs, Second Edition.* Lewis Publishers, Boca Raton, FL.

This book starts with a general overview of limnological processes and continues with a discussion of how to conduct a diagnostic analysis to determine which restoration techniques should be implemented in your reservoir. The book contains information on most lake restoration techniques including in-stream treatment, copper sulfate, biomanipulation, detention basins, destratification, and sediment removal. Detailed case studies are also provided.

Crook, J., G.F. Shimp, and J. Sandino (1996). Water Reuse Strategy for Santiago, Chile. *Proc. Joint AWWA/WEF Water Reuse Conference*, San Diego, CA.

Chile did not have regulations or standards for reuse of treated wastewater. This paper contains a review of current reuse standards from around the world and includes suggestions for water reuse standards that would be appropriate for Chile.

Cuff, Denis (1993). Oxygen Pump Aims to Save Reservoir Fish in Stale Water. *Contra Costa Times* June 22.

In this newspaper article, the East Bay Municipal Utility District's attempt to protect fish from stale water in the Comanche Reservoir by aerating the lake is discussed. Fish-suffocating water-oxygen depletion will be avoided using a 23' high Speece Cone oxygenator. The reservoir will be aerated for two years to test the effectiveness of Speece Cones in deep reservoirs.

Cullen, P., and C. Forsberg (1988). Experiences with Reducing Point Sources of Phosphorus to Lakes. *Hydrobiologia* 170:321–36.

The effects on lake water quality to reductions in external phosphorus loading are evaluated. The authors defined three types of responses including a change in trophic state, changes in phosphorus and chlorophyll *a* concentrations but not trophic state, and small or no change in lake water quality. Results showed that the percent reduction in in-lake concentrations was less than the percent reduction in inflow concentrations.

Danish Hydraulic Institute (2000a). *MIKE 11 Short Description*. Danish Hydraulic Institute, Horsholm, Denmark.

The MIKE 11 model is a modeling system for simulating flow, sediment transport, and water quality in rivers, reservoirs, and estuaries. The general model features, model processes, and model components are described in the document. The model simulates hydrology, hydrodynamics, advection and dispersion, sediment processes, and water quality.

Danish Hydraulic Institute (2000b). *MIKE 21 Environmental Modelling—A Short Description.* Danish Hydraulic Institute, Horsholm, Denmark.

The MIKE 21 model is a modeling system for simulating flow, sediment transport, and water quality in lakes, estuaries, and bays where stratification can be neglected. The general model features, model processes, and model components are described in this document. The model simulates hydrology, hydrodynamics, advection and dispersion, sediment processes, water quality, and eutrophication.

Danish Hydraulic Institute (2000c). *MIKE 3, a Three-dimensional Hydrodynamic Model, A Short Description*. Danish Hydraulic Institute, Horsholm, Denmark.

The MIKE 3 model is a three-dimensional hydrodynamic modeling system that includes three basic modules: the hydrodynamic module, the turbulence module, and the advection-dispersion module. Application modules include the water quality module and the eutrophication module. The general model features, model processes, and model components are described in this document.

Delfino, J.J., and G.F. Lee (1971). Variation of Manganese, Dissolved Oxygen and Related Chemical Parameters in the Bottom Waters of Lake Mendota, Wisconsin. *Water Research* 5:1207.

The variation of manganese and other related chemical parameters and their relation to oxygen levels at the water-sediment interface documented at Lake Mendota, Wisconsin, are discussed.

Dunst, R.C. (1981). Dredging Activities in Wisconsin's Lake Renewal Program, In: *Restoration of Lake and Inland Waters: International Symposium on Inland Waters and Lake Restoration,* EPA 440/5-81-010.

This reference provides general information on 12 dredging projects included in Wisconsin's dredging program for lake renewal. Information includes the size of the lakes, amount of sediment removed, costs, and disposal options used. More detailed information is provided for the Lilly Lake project, including data on sediment characteristics and water quality measurements before, during, and after dredging.

Dunst, R.C., J.G. Vennie, R.B. Corey, and A.E. Peterson. (1984). *Effect of Dredging Lilly Lake, Wisconsin.* EPA 600/3-84-097.

This document is a detailed report of the Lilly Lake Dredging Project. The study began in July 1976 and continued through the summer of 1982. Dredging occurred during 1978 and 1979. Lilly Lake had been infilling at a rate of 0.5 cm/yr and had accumulated up to 10.7 m of light weight organic sediments. Use of the lake was severely limited. The investigation included an assessment of water quality, aquatic biology, sediments, and hydrology before, during, and after dredging.

Ernst, M.R., W. Frossard, and J.L. Mancini (1994). Two Eutrophication Models Make the Grade, *Water Environment & Technology*, November, pp. 15–16.

The predictive abilities of the BATHTUB and WASP4 models were compared in this paper. Model predictions of nutrients and chlorophyll *a* were within 20% of measured data for both models. Information on data requirements and the time required for calibration was also presented.

Fast, A.W. (1979). Artificial Aeration as a Lake Restoration Technique, in *Lake Restoration.* EPA 440/5-79-001, pp. 121–32.

Use of artificial aeration as a lake restoration technique is discussed in this paper. Both destratification and hypolimnetic systems are discussed. Effects on lake biology, nitrogen gas formation, and the decision-making process are addressed.

Ferrara, T.A., and A. Hildick-Smith (1982). A Modeling Approach for Storm Water Quantity and Quality Control via Detention Basins. *Water Resources Bulletin* 18:975–81.

A model that can be used to design detention basins to control stormwater quantity and quality is presented in this paper. Information presented includes settling velocities for various sized particles and removal rates for sediment, total nitrogen, total phosphorus, and chemical oxygen demand for different storm conditions.

Florida Department of Environmental Protection (1996). *Reuse of Reclaimed Water and Land Application*. Chapter 62-610, Florida Administrative Code. FDEP, Tallahassee, FL.

Florida added a section on water reclamation in its administrative code in the late 1980s, which have since been revised in this document. Florida requires water reuse feasibility studies for wastewater treatment plants in areas where there are water supply problems. State legislation requires a reasonable amount of reclaimed water use unless the reuse is not feasible for technical, environmental or economic reasons.

Gambrell, R.P., C.N. Reddy, and R.A. Khalid (1983). Characterization of Trace and Toxic Materials in Sediments of a Lake Being Restored. *J. Water Poll. Cont. Fed.* 55:1201-10.

Work performed to characterize sediments to determine appropriate disposal options are discussed. The sediments were being removed from a lake during a lake restoration project. The paper discusses the method used to collect and analyze sediments and presents results of these analyses.

Harvey MUDD College Engineering Clinic (1993). *Final Report to Irvine Ranch Water District, Reclaimed and Nonpotable Water System Analysis and Recommended Strategies to Water Quality Problems.* Claremont, CA. Engineering Clinic.

Evaluations of reclaimed and nonpotable water system water quality problems and recommendation of alternatives including in-reservoir and post-reservoir treatment are presented.

Hayes, C.R., R.G. Clark, R.F. Stent, and C.J. Redshaw (1984). The Control of Algae by Chemical Treatment in a Eutrophic Water Supply Reservoir. *J. Inst. Water Eng. Sci* 38:149–62.

Measures taken at the Foxcote Reservoir including an application of ferric sulfate to the reservoir and chemical treatment of the pumped river inputs are documented. The treatment effectively reduced the phosphorus concentrations in inflow, but phosphorus recycling from bottom sediments was sufficient to cause algal blooms. After three years of treatment, the water quality has improved, perhaps due to the development of macrophyte beds.

Herbel, C.H. (1982). Grazing Management on Rangelands. *Journal of Soil and Water Conservation* 37(2):77–79.

The use of incorporating ecological principles into grazing management is emphasized in this paper. Specific principles discussed include proper utilization of the available forage plants, grazing intensity and rotation, stocking rate, animal distribution, proper plants, seed production of the perennial plants, flexibility, and an overall management plan.

Hernandez, P., R.B. Ambrose Jr., D. Prats, E. Ferrandis, and J.C. Asensi (1997). Modeling Eutrophication Kinetics in Reservoir Microcosms. *Water Research* 31(10):2511–19.

A model developed for the Amadorio Reservoir of Spain using WASP5 is described. An analysis was performed on the ability of WASP5 to handle daily phytoplankton and nutrient dynamics in perturbed microcosms. Four interacting systems were analyzed, including phytoplankton kinetics, the phosphorus cycle, the nitrogen cycle, and the balance of dissolved oxygen.

Herzfeld, M., and D. Hamilton (2000a). *The CWR Computational Aquatic Ecosystem Dynamics Model, CAEDYM, Science Manual*. Center for Water Research, The University of Western Australia, September 25, 2000. http://www.cwr.uwa.edu.au/~antenucc/cd/cwrsoft/doc/caedym_user/ title.html

CAEDYM is an ecological model designed to link to a hydrodynamic model. Seven major state variables are included in the CAEDYN model. The scientific basis for the simulations made in the model is described.

Herzfeld, M., and D. Hamilton, (2000b). *The CWR Computational Aquatic Ecosystem Dynamics Model, CAEDYM, User's Manual*. Center for Water Research, The University of Western Australia, September 25, 2000.
http://www.cwr.uwa.edu.au/~antenucc/cd/cwrsoft/doc/caedym_user/title.html

This document is the user's manual for the CAEDYM model. The CAEDYM is a library of subroutines executed through a "driver" hydrodynamic model. The user can define which biological groups will be simulated and the time steps to be used.

Hutchinson, G. E. (1957). *A Treatise on Limnology*. Vol. 1. Geography, Physics and Chemistry. John Wiley and Sons, New York, NY.

A comprehensive treatise on limnology focusing on the physical and chemical characteristics of lakes and water bodies is presented.

Hutchinson, G. E., and H. Lofler (1956). The Thermal Classification of Lakes. *Proceedings National Academy of Science* 42.

A comparison and classification of thermal lake types by latitude and altitude focusing on the number of periods of complete circulation in the year including amictic, cold monomictic, dimictic, warm monomictic, oligomictic, warm polymictic, and cold polymictic are provided.

Irvine Ranch Water District and Black & Veatch (1994). Final Report Watershed Management Study for Rattlesnake and Sand Canyon Reservoirs, Irvine, CA.

Case studies of two open storage reservoirs in a Southern California reclaimed water system are presented. Water quality is evaluated and assessed and water quality goals defined. Alternative management strategies are identified, screened, and the most feasible strategies more fully defined. Recommended in-reservoir and management strategies are then developed.

Kennedy, R.H. (1995). *Application of the BATHTUB Model to Selected Southeastern Reservoirs*. Technical Report EL-95-14, U.S. Army Engineer Waterways Experiment Station, Vicksburg, MS.

Application of the empirical BATHTUB Model to four U.S. Army Corps of Engineers constructed impoundments of the Chattahoochee and Coosa river basins of the southeastern U.S. is discussed. The reservoirs in this study are in basins that have experienced extensive development, with an increased risk of eutrophication and other water quality concerns in the lakes. The Bathtub model is used to predict eutrophication responses of the lakes to a variety of conditions.

Landers, J.C., and B.A. Knuth (1991). Use of Wetlands for Water Quality Improvement under the U.S. EPA Region V Clean Lakes Program. *Environ. Manage*. 15:151–62.

The use of wetlands at seven lakes that were part of the Clean Lakes Program for EPA Region V is examined in this paper. The authors reviewed management practices as well as ecological aspects of the projects. They stated that nutrients were reduced at each of the operating sites although specific data was not presented. They concluded that monitoring programs are generally inadequate to assess wetland performance.

Lang, G.A., and T.D. Fontaine (1990). Modeling the Fate and Transport of Organic Contaminants in Lake St. Clair. *Journal of Great Lakes Research* 16(2):216–32.

A model of Lake St. Clair developed to predict and understand the transport and fate of organic contaminants using a multi-segmented contaminant mass balance model based upon EPA's TOXIWASP model is described. TOXIWASP combines the kinetic formulations in Exposure Analysis Modeling System (EXAMS) and the transport processes in the Water Analysis Simulation Program (WASP). Model applications included simulations of octachlorostyrene (OCS) and polychlorinated biphenyl (PCB) dynamics of the lake.

Lowe, E.F., L.E. Battoe, D.O. Stites, and M.F. Coveney (1992). Particulate phosphorus removal via wetland filtration: an examination of potential for hypertrophic lake restoration. *Environ. Manage*. 16:67–74.

An evaluation of phosphorus removal using wetland filtration and its effect on a large hypereutrophic lake in Florida is provided. Expected phosphorus concentrations were predicted for various scenarios of external loading rates.

Lung, W., and C.E. Larson (1995). Water Quality Modeling of the Upper Mississippi River and Lake Pepin. *Journal of Environmental Engineering* 121 (10):691–99.

WASP5 includes two submodels for water quality, one for eutrophication and one for toxics. An application of the EUTRO5 submodel is discussed in this paper. Several scenarios for reducing phosphorus loading to the upper Mississippi River and Lake Pepin were evaluated, and results are presented.

Lusby, G.C. (1989). *Effects of Grazing on Runoff and Sediment Yield from Dessert Rangeland at Badger Wash in Western Colorado, 1953–1973*. USES Water Supply Paper 1532-1, Washington, D.C.

Results of a 20-year study to investigate the differences in runoff and sediment yield between grazed and ungrazed watersheds are presented in this report. The study area, in western Colorado, receives about 9" of rain per year and is classified as arid to semiarid. Ungrazed areas accounted for 60% of the runoff and 37% of the sediment being produced from grazed areas.

Lynch, S., K. Martin, and D. Bramwell (1998). Southern California Comprehensive Water Reclamation and Reuse Study. *Proc. WEF Watershed Specialty Conference,* Denver, CO, 1227.

The first phase of a study to identify a regional reclamation system for Southern California and develop capital projects is discussed. The study included development of a regional database for reclaimed water and an evaluation of increasing usage through groundwater recharge, indirect potable use, and environmental enhancement.

Martin, E.H. (1988) Effectiveness of an Urban Runoff Detention Pond Wetland System. *Journal of Environmental Engineering* 114 (3):810–27.

The effectiveness of a detention pond and wetland system is documented. Removal efficiencies are measured for the detention pond and wetland for major ions, metals, and nutrients.

McGinn, J., L. Barbato, J. Van Teylingen, and V. L. Card (1998). Colorado Springs: A Model for Reclaimed Water Systems. *Proc. Joint AWWA/WEF Water Reuse Conference*, Lake Buena Vista, FL, 497.

The water reuse system for Colorado Springs, CO is described. Issues discussed include the demand and location of customers; desired pressure; the need for storage to meet peak demands, operation and maintenance concerns, water quality considerations, and financial aspects of the reuse system.

McKnight, D.M., S.W. Chisholm, and D.R.F. Harleman (1983). CuSO4 Treatment of Nuisance Algal Blooms in Drinking Water Reservoirs. *Environ. Manage.* 7:311–20.

Toxicity of copper in various forms to algal species and the processes affecting copper speciation in reservoirs is discussed. Chemical processes such as precipitation, adsorption to suspended sediments, and complexation by organic substances are important factors in controlling chemical speciation. Dosage requirements can be more effectively set by considering specific water chemistry and algal composition found in the reservoir.

Meyer, B. (1991). Pneumatic Destratification System Design Using a Spreadsheet Program. *Information Exchange Bulletin,* Vol. E-91-1, WOTS U.S. Army Corps Engineers, Waterways Experiment Station, Vicksburg, MS.

A spreadsheet program available from the Army Corps of Engineers that can be used to design a compressed air destratification system is described. Required inputs include surface area of the lake, volume-capacity curve, typical temperature profiles, depth of water above the diffuser, time period desired to achieve destratification, internal diameter of the diffuser line, internal diameter of the supply line, and an approximate length of the supply line.

Mortimer, C.H. (1941). The Exchange of Dissolved Substances between Mud and Water in Lakes. 1 and 2. *J. Ecol.* 29:280–329.

The first half of the results of extensive research on the interaction between mud and water in lakes are presented.

Mortimer, C.H. (1942). The Exchange of Dissolved Substances between Mud and Water in Lakes. 3 and 4. *J. Ecol.* 30:147–201.

The second half of the results of extensive research on the interaction of mud and water in lakes are presented.

Mortimer, C.H. (1971). Chemical Exchanges Between Sediments and Water in the Great Lakes-Speculations on Probable Regulatory Mechanisms. *Limnol. Oceanography* 16:387–404.

Interaction between mud and water in lakes reported previously is updated, and previous research is expanded upon.

National Research Council (1998). *Issues in Potable Reuse*. National Academy Press, Washington, D.C.

An NRC committee concluded that reclaimed wastewater can be used to supplement drinking-water sources, but only as a last resort and after a thorough health and safety evaluation. Municipalities first must fully assess health impacts from likely contaminants and develop comprehensive systems for monitoring, testing, and treatment. Other water resources and conservation measures also should be tried to the extent practical, before turning to reclaimed wastewater. Because regulations for safe drinking water were not developed with reclaimed water in mind, they may not be the best standard for testing its quality. Reclaimed water may contain sources of contamination that cannot be determined through current testing of treatment processes.

Nichols, D.S. (1983). Capacity of Natural Wetlands to Remove Nutrients from Wastewater. *Journal WPCF* 55:495–505.

Mechanisms by which phosphorus and nitrogen are removed from wastewater by wetlands are reviewed, and the relationship between nutrient application rates and removal by wetlands are described.

Noble, R.L. (1986). Predator-prey Interactions in Reservoir Communities. In: *Reservoir Fisheries Management: Strategies for the 80's.* G.E. Hall and M.J. Van Den Avyle (eds.), Reservoir Comm. SDAFS, Bethesda, MD.

Predator-prey interactions that occur among several trophic levels are discussed. The article emphasizes that management decisions are often made on adult fish behavior and that other factors are often overlooked. These other factors include eating behaviors at different life stages, prey density and availability, the timing of algal blooms and fish spawning, macrophyte distribution, and reservoir operations including water level changes. To manage reservoir fisheries effectively, the entire ecosystem at all trophic levels must be understood and considered in management decisions.

Oberts, G.L., and R.A. Osgood (1991). Water-quality Effectiveness of a Detention/Wetland Treatment System and Its Effect on an Urban Lake. *Environ. Manage.* 15:131–38.

The results of a study of a detention basin wetlands system in Minnesota that receives stormwater runoff are presented. Although a 75% reduction in total phosphorus was measured in the lake, concentrations of phosphorus were unchanged due to internal nutrient loading.

Olem, H., and G. Flock, eds. (1990). *Lake and Reservoir Restoration Guidance Manual, 2nd edition.* EPA 440/4-90-006. Prepared by North American Lake Management Society for U.S. Environmental Protection Agency, Washington, D.C.

A revision of the 1988 *Guidance Manual*, information on most lake restoration techniques including six related to the project: dredging, destratification, copper sulfate, oxygenation, biomanipulation, and detention basins are contained in this manual. A few case studies are included. The information is more general in nature than that presented in Cooke et al. (1993).

Owens, L.B., W.M. Edwards, and R.W. Van Keuren (1989). Sediment and Nutrient Losses from an Unimproved All-year Grazed Watershed. *J. Environ. Qual.* 18, p.232.

A case study of an unimproved pasture in east-central Ohio, which was used for year-round grazing for six years and was ungrazed for two years is presented. Sediment yield was nine times higher during the year-round grazing. Runoff increased by 10%.

Palmer, M. (2001). *Water Quality Modeling*. The World Bank, Washington, D.C.

A range of water quality models is discussed in this book. The document includes many examples of water quality modeling used in World Bank projects throughout the world. Topics discussed in the book include water quality model structure and processes, descriptions of commonly used models, and case studies.

Parker, Sybil P. (1984). *Concise Encyclopedia of Science and Technology*. McGraw-Hill Book Company, New York, NY, p.354.

A concise summary of the different types of chlorophyll and in what water bodies they can be found is provided.

Pastorak, R.A., M.W. Lorenzen, and T.C. Ginn (1982). *Environmental Aspects of Artificial Aeration/and Oxygenation of Reservoirs: A Review of Theory, Techniques, and Experiences.* Technical Report No. E-82-3, U.S. Army Corps Engineers Waterways Exp. Sta., Vicksburg, MS.

A review and evaluation of the effects of destratification and hypolimnetic aeration on lake water quality is provided in this document. The results of many lake studies are summarized and presented. Effects of water quality, plankton, macroinvertebrates, and fish are discussed.

Peterson, S.A. (1982). Lake Restoration by Sediment Removal. *Water Resources Bulletin* 18:423–35.

Information on the effectiveness of sediment removal as a lake restoration technique is reviewed and updated. The purposes of and considerations for sediment removal, including characterization of the sediment, the depth of sediment removal, environmental problems associated with dredging, removal method, disposal areas, and costs are discussed. Five case studies are presented.

Prato, T., and H. Shi (1990). A Comparison of Erosion and Water Pollution Control Strategies for an Agricultural Watershed. *Water Resources Research* 26(2):199–205.

Two erosion control strategies with a pollution control strategy using a riparian buffer are compared. The two erosion control strategies caused less total erosion and sediment and nutrient pollution than the riparian strategy.

Raman, R.K., and R.L. Evans (1984). *Aeration-destratification of Lake Eureka: Second Year Operation.* ISWS/CIR-159/84, Circular 159, State of Illinois, Department of Energy and Natural Resources.

The results of the second year of operation of a destratification system on a small water supply lake in Illinois are presented. The system kept the lake destratified, and significant reductions in lake concentrations of nutrients were measured.

Rausch, D.L., and J.D. Schreiber (1981). Sediment and Nutrient Trap Efficiency of a Small Flood-Detention Reservoir. *Journal of Environmental Quality* 10(3):288–93.

The results of a three-year study of a small flood retention reservoir in central Missouri are presented. The purpose of the study was to compare the trap efficiencies of sediment and nutrients and determine the factors that affected them. Trap efficiencies are presented for sediment, sediment P, and solution P.

Reckhow, K.H. (1990). *EUTROMOD Spreadsheet Program—A Regional Modeling Scheme for Nutrient Runoff and Lake Trophic State Modeling.* School of Forestry and Environmental Studies, Duke University, Durham, NC.

The EUTROMOD spreadsheet program is described in this document. Required model inputs, limitations, and model outputs are discussed.

Robbins, R.W., J.L. Glicker, D.M. Bloem, and B.M. Niss (1991). *Effective Watershed Management for Surface Water Supplies.* AWWA Research Foundation, Denver, CO, 401p.

This text includes discussions of developing watershed management plans and descriptions of structural and nonstructural BMPs.

Rodhe, W. (1948) Environmental Requirements of Freshwater Plankton Algae. Experimental Studies in the Ecology of Phytoplankton. *Symbol. Bot. Upsalien* 10 (1), 149 pp.

A thorough summary of minimum, optimum, and maximum primary nutrient requirements and the impacts on plankton algae growth and selection is presented.

Sartoris, J.J., and J.T. Boehmke (1987) *Limnological Effects of Artificial Aeration at Lake Cachuma, California, 1980-1984.* REC-ERC-87-10, U.S. Department of the Interior, Bureau of Reclamation.

An evaluation of an aeration system installed on Lake Cachuma, California is presented. Although the system never fully destratified, the lake, positive water quality benefits were observed. Data on changes in temperature, DO redox potential, light penetration, metals, nutrients, chlorophyll *a*, plankton, and benthic organisms are presented.

Schladow, S.G. (1993). Lake Destratification by Bubble-Plume Systems: Design Methodology. *Journal of Hydraulic Engineering* 119 (3):350–68.

A new design method for bubble-plume destratification systems is presented. The method can be used to calculate the required airflow rate and number of buoyancy sources. The method was verified by simulating existing destratification systems. Three systems designed with the new method were being installed at the time of publication.

Schueler, T.R. (1987). *Controlling Urban Runoff: A Practical Manual for Planning and Designing Urban BMPs.* Washington, D.C., Metropolitan Washington Council of Governments.

Information on several structural BMPs, including detention basins. Information on design criteria and characteristics and maintenance is provided.

Shoemaker, L., M. Lahlou, M. Bryer, D. Kumar, and K. Kratt (1997). *Compendium of Tools for Watershed Assessment and TMDL Development.* EPA841-B-97-006. U.S. Environmental Protection Agency, Office of Water, Washington, D.C.

A wide range of modeling tools is summarized in this report. Three types of models are presented: watershed loading models, receiving water models, and ecological assessment techniques. In addition to information of specific models, the report contains a comparative analysis of various models and guidance on model selection.

Spangenberg, C. (1991). *Assessment of Hypolinmetic Aeration as A Restorative Technique for Irvine Lake.* Thesis for California State University, Long Beach.

An evaluation of the water quality of a freshwater reservoir in Southern California and alternative technologies for promoting freshwater fishery habitat while maintaining temperature profile are presented in this thesis.

Speece, Richard E. (1994). Lateral Thinking Solves Stratification Problems. *WQI* 3:12–15.

Adding commercial oxygen to the hypolimnion of stratified lakes and impoundments can help remedy water quality deterioration. The advantages and design of horizontal-flow aeration systems are discussed. Hypolimnion oxygenation versus destratification and the use of commercial oxygen compared to air are also addressed.

Speece, R.E. (1996). Oxygen Supplementation by U-Tube to the Tombigbee River. *Wat. Sci. Tech.* 34(12):83–90.

Because the effluent from two Alabama paper mills was high in biochemical oxygen demand (BOD), it had a noticeable effect on the Tombigbee River's dissolved oxygen (DO) concentration. To avoid depletion of the river's DO, the mills were permitted to supplement DO to their effluent to offset its BOD. Selection and operation of the oxygen supplementation systems installed at each mill are presented and discussed.

Speece, R.E. (1989). *Tailwaters: A Regulated Streams' Most Valuable Resource*. Prepared for North American Lake Management Society Plenary Session.

The use of dams to regulate tailwater streamflow can create previously unavailable or unreliable recreational opportunities. Impounding a free-flowing river can alter discharge pattern, water temperature, and water DO. Each of these can directly impact recreational activities such as swimming, rafting, and fishing. Case histories and operational strategies aimed at maximizing the public benefit of streams controlled by dam discharge are discussed.

Speece, R.E. (2000). *Proposed Design of an Oxygen Supplementation System for Percy Priest Reservoir*. Prepared for the Nashville U.S. Army Corp of Engineers.

Adding commercial oxygen to the hypolimnion of stratified lakes and impoundments can help remedy water quality deterioration. The advantages of using commercial oxygen in horizontal-flow aeration systems are reviewed. The proposed aeration system design for Percy Priest Reservoir is also discussed.

Speece, R.E., N. Nirmalakhandan, and G. Tchobanoglous (1990). Commercial Oxygen Use in Water-Quality Management. *Water Environment & Technology* July:54–61.

Use of commercial oxygen for the water-quality management of wastewater treatment, lake management, and hydropower generation is discussed. The article reviews commercial oxygen aeration equipment, oxygen generation equipment and processes, and costs. Case studies of full-scale commercial applications are also included.

Striegl, R.G. (1987). Suspended Sediment and Metals Removal from Urban Runoff by a Small Lake. *Water Resources Bulletin* 23(6):985–996.

A study conducted on a small lake in the Chicago Metropolitan area to determine the lake's efficiency in removing suspended sediments and metals from urban runoff is presented and discussed. The results showed that the lake was from 91 to 95% efficient in removing sediments and from 76 to 94% efficient in removing copper, lead, iron, and zinc.

Stumm, W., and J. J. Morgan (1970). *Aquatic Chemistry*. Wiley Interscience, New York, NY.

This definitive textbook on water chemistry provides general principles of natural water chemistry including chemical thermodynamics, solid-solution interface and kinetics, kinetics of redox processes, photochemical processes, atmospheric-water interactions, precipitation and dissolution, particle interaction, oxidation and reduction, microbial mediation, equilibrium, and composition of natural waters.

Tchobanogolous, G., and E.D. Schroeder (1985). *Water Quality*. Addison-Wesley, USA. pp. 121–23, 188–89.

This document contains a comprehensive examination of water quantity and water quality, analytical methods for water quality management, modeling of water quality and modification of water quality. Sections are provided on chemical characteristics of gases (solubility, gas transfer) and problems associated with biological organisms in water storage facilities referenced.

Tetra Tech, Inc. (1995). *Watershed Protection: Clean Lakes Case Study, Phosphorus Inactivation and Wetland Manipulation, Kezar Lake, NH*. EPA 841-F-95-002

Lake restoration work conducted at Kezar Lake is summarized. Sediment phosphorus inactivation and wetlands management were implemented to address water quality problems. Results of post-restoration monitoring, conducted over a 10-year period, are summarized in the case study report.

Thomann, R.V., and J.A. Mueller (1987). *Principles of Surface Water Quality Modeling and Control*. Harper & Row, New York, NY.

A wide range of surface water models and modeling issues including a detailed derivation of the relationship between BOD, DO, and nutrients is described in this textbook.

Trott, K.E., and M.J. Singer (1983). Relative Erodibility of 20 California Range and Forest Soils. *Journal of Soil Science of America* 47:753–59.

Under laboratory conditions, the relative erodibilities of 20 upland California soils were determined and compared to values predicted from the Universal Soil Loss Equation. For the soils studied, mineralogy was an important factor in predicting erodibility. The authors suggest that soil color and bulk soil mineralogy be used in addition to the USLE to determine erodibility.

Tufford, D.L., and H.N. McKeller (1999). Spatial and Temporal Hydrodynamic and Water Quality Modeling Analysis of a Large Reservoir on the South Carolina Coastal Plain. *Ecological Modeling* 114.

A two-dimensional, 31-segment, 61-channel hydrodynamic and water quality model of Lake Marion, SC using WASP5 is described. Phytoplankton kinetic rates and constants were obtained from field studies of Lake Marion and existing modeling literature. The presence of additional nutrient processes not conceptualized in model mechanisms was suggested. Implementation of the WASP5 model to Lake Marion suggests "ecological variability in the lake due to constraints on the specification of rate constants."

Uhlmann, D., and H. Klapper (1985). Protection and Restoration of Lakes and Reservoirs in the German Democratic Republic—Case Studies. *Conference Proceedings from Lake Pollution and Recovery*: European Water Pollution Control Association International Congress, Rome, Italy, pp. 262–71.

Various methods used to control nonpoint external sources of nutrients to reservoirs ranging from eutrophic to oligotrophic along with internal measures, interrelationships to these approaches and biological responses; dynamic "SALMO" model used to predict ecological benefits, and drawbacks of measures are described.

U.S. Army Corps of Engineers (1982). *CE-QUAL-R1: A Numerical One-Dimensional Model of Reservoir Water Quality; User's Manual*. Instruction Report E-82-1 revised edition. U.S. Army Engineer Waterways Experiment Station Environmental Laboratory, Vicksburg, MS, 623 pp.

The users manual for the CE-QUAL-R1 model includes four main sections. An overview of the model is provided in the first section. Model capabilities and limitations and the basis equations used in the model are contained in the second section. The third section provides detailed information on each of the model components. How to select coefficients and constants and how to calibrate the model are considered in the fourth section. CE-QUAL-R1 is a one-dimensional model (horizontally averaged) that provides a vertical distribution of reservoir water quality.

U.S. Army Corps of Engineers (2000) *CE-QUAL-R1, One-dimensional (Vertical) Reservoir Water Quality Model*. Environmental Laboratory, Vicksburg, MS. http://www.wes.army.mil/el.elmodels/r1info.html

CE-QUAL-R1 is a one-dimensional (vertical) reservoir water quality model. The capabilities and operation of the CE-QUAL-R1 model are documented.

U.S. EPA (1984). *Technical Guidance Manual for Performing Waste Load Allocations – Book IV Lakes and Impoundments—Chapter 2 Nurtient/Eutrophication Impacts*. EPA 440/4-84-019, U.S. Environmental Protection Agency, Office of Water Regulations and Standards. Washington, D.C.

Lake eutrophication processes and the factors that affect waste load allocations in lakes are discussed in this document. Three class of models that could be useful in performing lake evaluations are also discussed.

U.S. EPA (1986). *Quality Criteria for Water, 1986.* PB87-226759 U.S. Environmental Protection Agency, Washington, D.C.

Section 304(a)(1) of the Clean Water Act requires EPA to publish and periodically update ambient water quality criteria. These criteria present scientific data and guidance of the environmental effects of pollutants, which can be useful to derive regulatory requirements based on considerations of water quality impacts. Periodically EPA and its predecessor agencies has issued water quality criteria, beginning in 1968 with the "Green Book" followed by the 1973 publication of the "Blue Book" (*Water Quality Criteria 1972*). In 1976 the "Red Book" (Quality Criteria for Water) was published. In a continuing effort to provide up-to-date criteria and associated information, this document, the "Gold Book" (*Quality Criteria for Water 1986*), was prepared.

U.S. EPA (1988). *Design Manual, Constructed Wetlands and Aquatic Plant Systems for Municipal Wastewater Treatment*. EPA/625/1-88/022. U.S. Environmental Protection Agency, Washington, D.C.

Aquatic treatment systems are reviewed and evaluated with regards to environmental and public health considerations, including nutrients, pathogens, metals, and trace organics. Design requirements for both constructed wetlands and aquatic plant systems are described in detail.

U.S. EPA (1991). *Municipal Water Reuse, Selected Readings on Water Reuse.* EPA 430/9-91-022. U.S. Environmental Protection Agency, Washington, D.C.

This publication contains a series of articles on water reuse, reprinted from Water Pollution Control Federation's *Water Environment & Technology Journal* in 1990 and 1991.

U.S. EPA (1992). *Guidelines for Water Reuse*. EPA/625/R-92/004. U.S.EPA, Center for Environmental Research Information, Cincinnati, OH.

EPA's guidance manual on reuse is not a formal regulatory document, but does provide recommendations for a wide range of reuse practices. Many reuse issues including types of reuse applications and water reuse regulations are addressed in this document. The guidelines provide general recommendations to indicate the type of treatment and water quality criteria likely to be required.

U.S. EPA (1995). *Watershed Protection: Clean Lakes Case Study, Watershed and In-Lake Practices Improve Green Valley Lake, Iowa*. EPA 841-F-95-003.

A Clean Lakes project was conducted at Green Valley Lake, Iowa, and results are summarized in this report. BMPs included improved agricultural management and construction of sediment/nutrient retention dikes in the two major arms of the lake. Results of post-restoration monitoring are also included.

Walker, W.W. (1987). *Empirical Methods for Predicting Eutrophication in Impoundments.* Report 4, Phase III, Application Manual. Technical Report E-81-9. U.S. Army Corps of Engineers, Vicksburg, MS.

Model documentation for the BATHTUB model is contained in this reference. The BATHTUB model includes nutrient balance models and eutrophication response models in a spatially segmented reservoir network. BATHTUB is an empirical model that accounts for the processes of advection, diffusion and nutrient sedimentation. FLUX and PROFILE, two models that assist with data reduction and analysis, are also included.

Walker, W.W. (1996). *Simplified Procedures for Eutrophication Assessment and Prediction: User Manual.* Instruction Report W-96-2. U.S. Army Corps of Engineers, Waterways Experiment Station, Water Operations Technical Support Program.

This document is the updated User's Manual for the BATHTUB model. Walker (1985) is updated, and information on the interface for the PC version of the model is presented.

Washington State Department of Health (1997). *Water Reclamation and Reuse Standards.* WDOH, Olympia, WA.

The State of Washington recently revised its Water Reclamation and Reuse Standards. Reclaimed water may be used as a source of supply for landscape impoundments, restricted recreational impoundments, and nonrestricted recreational impoundments. The state has not developed provisions for use of reclaimed water in augmenting surface waters for indirect potable reuse.

Wedepohl, R.E., D.R. Knauer, G.B. Wolbert, H. Olem, P.J. Garrison, and K. Kepford (1990). *Monitoring Lake and Reservoir Restoration.* EPA 440/4-90-007. Prepared by N. Am. Lake Manage. Soc. for U.S. Environmental Protection Agency, Office of Water, Washington, D.C.

This document is a technical supplement to The Lake and Reservoir Restoration Guidance Manual. Additional information specifically related to developing and conducting monitoring programs for lake restoration projects is presented in this report.

Wen, J., and R.S. Torrest (1987). Aeration-Induced Circulation from Line Sources. I: Channel Flows. *J. of Environmental Engineering* 113 (1):82–98.

Experimental studies on the fluid mechanics of aeration-induced circulation are described. Detailed measurements of velocity profiles are presented for a wide range of aeration rates. The circulation cell was found to be about four times the water depth.

Wen, J., and R.S. Torrest (1987). Aeration-Induced Circulation from Line Sources. II: Dissolved Oxygen Variations. *J. of Environmental Engineering* 113 (1):99–105.

Companion paper to the one listed above. Data on the buildup of dissolved oxygen from aeration systems are presented. A model is presented that describes the DO build-up.

Wetzel, R. (1975). *Limnology.* W. B. Saunders Company, Philadelphia, PA, pp. 78–79, 215–45.

The physical, chemical, and biological properties and mechanisms that occur in limnology are described in this book.

Wetzel, R. (1983). *Limnology*. 2nd ed. Saunders. Philadelphia, PA.

This book describes physical, chemical and biological properties and mechanisms that occur in limnology.

APPENDIX B

GRAY LITERATURE SURVEY INFORMATION

IMPACT OF SURFACE STORAGE ON RECLAIMED WATER: SEASONAL & LONG-TERM
WERF Project No. 99-PUM-4

CONTACT & RESPONSE LIST

Agency	Reponded (Y/N)	Surface Storage (Y/N)
Arcadus, Geraghty, & Miller	N	
Austin (City of) Water Dept.	N	
Buena Sanitation District	Y	Y
Dublin San Ramon Services District	Y	N
Eastern Municipal Water District	Y	Y
Irvine Ranch Water District	Y	Y
Las Virgenes Municipal Water District	Y	Y
Los Angeles (County of) Sanitation District	Y	N
Monterey Regional Water Poll. Control	Y	Y
Orange County Public Utilities	Y	N
Orlando (City of)	N	
Phoenix (City of) Water Service Dept.	Y	N
Pinellas County Utilities	Y	Y
Santa Clara Valley Water District	Y	N
Santa Margarita Water District	Y	Y
South Bay Water Recycling	Y	Y
Tucson Water Dept.	Y	Y
Vallecitos Water District	Y	Y
Vernon (City of), British Columbia	Y	Y
Westminister (City of)	Y	N

IMPACT OF SURFACE STORAGE ON RECLAIMED WATER: SEASONAL & LONG-TERM
WERF RFP NO. 99-PUM-4

Survey Reponse for: *Buena Sanitation District (Shadowridge Water Reclamation Plant)*
Information Received: *05/02/00*

Water quality management strategies used:

- [ ] Destratification
- [ ] Aeration
- [ ] Hypolimnetic Withdrawal
- [x] Dilution
- [ ] Other Chemical Trtmt
- [ ] Dredging
- [ ] Wetlands
- [ ] Biomanipulation
- [ ] Copper Sulfate/Cuprine Treatment
- [ ] Alum Treatment
- [ ] Filtration
- [ ] Other (specify):

Properties of Open Storage Reservoir(s):

Reservoir Name	Capacity (acre-feet)	Surface Area (acres)	Annual Usage (acre-feet)	Maximum Depth (feet)	Mean Depth (feet)
Shadowridge	11		365	16	14

Water quality monitoring program conducted at:

- [x] Treatment Plant Effluent
- [ ] Distribution System
- [ ] Storage Reservoirs

Types of water quality parameters monitored for each location used above:

Parameters	Treatment Plant Effluent	Distribution System	Storage Reservoirs
Turbidity	X		
Suspended Solids	X		
EC			
pH	X		
Dissolved Oxygen			X
Temperature	X		
Algae Counts			
Algae Identification			
Chlorine Residual	X		
Color			
Total Coliform	X		
Phosphorus			
TKN			
Nitrate + Nitrite			
Silicate			
Others (list)			

Frequency of water quality sampling in the storage reservoir(s):

- [ ] Weekly
- [x] Monthly
- [ ] Quarterly
- [ ] In Response to Problems
- [ ] Other (specify):

Water quality problems experienced:

- [x] Algae blooms
- [ ] Odor
- [ ] Turbidity
- [ ] Other (specify):

IMPACT OF SURFACE STORAGE ON RECLAIMED WATER: SEASONAL & LONG-TERM
WERF RFP NO. 99-PUM-4

Survey Reponse for: *Eastern Municipal Water District*
Information Received: *04/05/00*

Water quality management strategies used:

- [ ] Destratification
- [ ] Aeration
- [ ] Hypolimnetic Withdrawal
- [x] Dilution
- [ ] Other Chemical Trtmt
- [ ] Dredging
- [ ] Wetlands
- [ ] Biomanipulation
- [ ] Copper Sulfate/Cuprine Treatment
- [ ] Alum Treatment
- [ ] Filtration
- [x] Other (specify): Maintain water depth > 3 feet

Properties of Open Storage Reservoir(s):

Reservoir Name	Capacity (acre-feet)	Surface Area (acres)	Annual Usage (acre-feet)	Maximum Depth (feet)	Mean Depth (feet)
Trumble Road	881	27.0			
Skiland	1,013	66.6			
Winchester Ponds	1,381	72.6			
MWD Ponds	227	??			
Alessandro Ponds	169	21.5			

Water quality monitoring program conducted at:

- [x] Treatment Plant Effluent
- [ ] Distribution System
- [x] Storage Reservoirs

Types of water quality parameters monitored for each location used above:

Parameters	Treatment Plant Effluent	Distribution System	Storage Reservoirs
Turbidity	X		
Suspended Solids	X		
EC	X		
pH	X		
Dissolved Oxygen			
Temperature			
Algae Counts			
Algae Identification			
Chlorine Residual	X		
Color			
Total Coliform	X		
Phosphorus	X		
TKN	X		X
Nitrate + Nitrite			
Silicate			
Others (list)			

Frequency of water quality sampling in the storage reservoir(s):

- [ ] Weekly
- [x] Monthly
- [ ] Quarterly
- [ ] In Response to Problems
- [x] Other (specify): Permit requirements

Water quality problems experienced:

- [x] Algae blooms: Occassional
- [ ] Odor
- [ ] Turbidity
- [ ] Other (specify):

IMPACT OF SURFACE STORAGE ON RECLAIMED WATER: SEASONAL & LONG-TERM
WERF RFP NO. 99-PUM-4

Survey Reponse for: *Irvine Ranch Water District*
Information Received: *04/04/00*

Water quality management strategies used:

- [x] Destratification
- [x] Aeration
- [x] Hypolimnetic Withdrawal
- [ ] Dilution
- [ ] Other Chemical Trtmt
- [ ] Dredging
- [ ] Wetlands
- [ ] Biomanipulation
- [ ] Copper Sulfate/Cuprine Treatment
- [ ] Alum Treatment
- [x] Filtration
- [ ] Other (specify):

Properties of Open Storage Reservoir(s):

Reservoir Name	Capacity (acre-feet)	Surface Area (acres)	Annual Usage (acre-feet)	Maximum Depth (feet)	Mean Depth (feet)
Sand Canyon	768	42	400	40	18
Rattlesnake	1440	60	900	45	24

Water quality monitoring program conducted at:

- [x] Treatment Plant Effluent
- [x] Distribution System
- [x] Storage Reservoirs

Types of water quality parameters monitored for each location used above:

Parameters	Treatment Plant Effluent	Distribution System	Storage Reservoirs
Turbidity	X	X	X
Suspended Solids	X	X	X
EC	X	X	X
pH	X	X	X
Dissolved Oxygen			X
Temperature			X
Algae Counts			X
Algae Identification			X
Chlorine Residual	X	X	X
Color	X	X	X
Total Coliform	X	X	X
Phosphorus	X		
TKN	X		
Nitrate + Nitrite	X		
Silicate			
Others (list)			

Frequency of water quality sampling in the storage reservoir(s):

- [x] Weekly
- [ ] Monthly
- [ ] Quarterly
- [x] In Response to Problems
- [ ] Other (specify):

Water quality problems experienced:

- [x] Algae blooms
- [x] Odor
- [x] Turbidity
- [x] Other (specify): Anoxic & anaerobic conditions

IMPACT OF SURFACE STORAGE ON RECLAIMED WATER: SEASONAL & LONG-TERM
WERF RFP NO. 99-PUM-4

Survey Reponse for: *Las Virgenes Municipal Water District*
Information Received: *03/22/00*

Water quality management strategies used:

- [ ] Destratification
- [ ] Aeration
- [ ] Hypolimnetic Withdrawal
- [ ] Dilution
- [ ] Other Chemical Trtmt
- [ ] Dredging
- [ ] Wetlands
- [ ] Biomanipulation
- [ ] Copper Sulfate/Cuprine Treatment
- [ ] Alum Treatment
- [ ] Filtration
- [x] Other (specify): Bypassing; Residual chlorine from treatment

Properties of Open Storage Reservoir(s):

Reservoir Name	Capacity (acre-feet)	Surface Area (acres)	Annual Usage (acre-feet)	Maximum Depth (feet)	Mean Depth (feet)
Reservoir No. 1	12	0.6		30	
Reservoir No. 2	45	2.7		27	
Reservoir No. 3	2	0.25		14	

Water quality monitoring program conducted at:

- [x] Treatment Plant Effluent
- [ ] Distribution System
- [ ] Storage Reservoirs

Types of water quality parameters monitored for each location used above:

Parameters	Treatment Plant Effluent	Distribution System	Storage Reservoirs
Turbidity	X		
Suspended Solids	X		
EC			
pH	X		
Dissolved Oxygen	X		
Temperature	X		
Algae Counts			
Algae Identification			
Chlorine Residual	X		
Color			
Total Coliform	X		
Phosphorus	X		
TKN	X		
Nitrate + Nitrite	X		
Silicate			
Others (list)	arsenic	selenium	
	barium	silver	
	cadmium	cyanide	
	chromium	fluoride	
	lead	radioactivity	
	mercury		

Frequency of water quality sampling in the storage reservoir(s):

- [ ] Weekly
- [ ] Monthly
- [ ] Quarterly
- [x] In Response to Problems
- [ ] Other (specify):

Water quality problems experienced:

- [x] Algae blooms
- [x] Odor
- [x] Turbidity
- [ ] Other (specify):

IMPACT OF SURFACE STORAGE ON RECLAIMED WATER: SEASONAL & LONG-TERM
WERF RFP NO. 99-PUM-4

Survey Reponse for: *Monterey Regional Water Pollution Control Agency*
Information Received: *03/28/00*

Water quality management strategies used:

- [ ] Destratification
- [x] Aeration
- [ ] Hypolimnetic Withdrawal
- [ ] Dilution
- [x] Other Chemical Trtmt
- [ ] Dredging
- [ ] Wetlands
- [ ] Biomanipulation
- [ ] Copper Sulfate/Cuprine Treatment
- [ ] Alum Treatment
- [ ] Filtration
- [ ] Other (specify):

Properties of Open Storage Reservoir(s):

Reservoir Name	Capacity (acre-feet)	Surface Area (acres)	Annual Usage (acre-feet)	Maximum Depth (feet)	Mean Depth (feet)
SVRP Pond	80	4		24	15

Water quality monitoring program conducted at:

- [x] Treatment Plant Effluent
- [x] Distribution System
- [x] Storage Reservoirs

Types of water quality parameters monitored for each location used above:

Parameters	Treatment Plant Effluent	Distribution System	Storage Reservoirs
Turbidity	X		
Suspended Solids	X		
EC	X		
pH	X		
Dissolved Oxygen			X
Temperature			
Algae Counts			X
Algae Identification			
Chlorine Residual	X	X	X
Color			
Total Coliform	X		X
Phosphorus			
TKN	X		
Nitrate + Nitrite	X		
Silicate			
Others (list)			

Frequency of water quality sampling in the storage reservoir(s):

- [x] Weekly
- [ ] Monthly
- [ ] Quarterly
- [x] In Response to Problems
- [ ] Other (specify):

Water quality problems experienced:

- [x] Algae blooms
- [ ] Odor
- [ ] Turbidity
- [x] Other (specify): Coliform spikes

IMPACT OF SURFACE STORAGE ON RECLAIMED WATER: SEASONAL & LONG-TERM
WERF RFP NO. 99-PUM-4

Survey Reponse for: ***Pinellas County Utilities***
Information Received: ***04/01/00***

Water quality management strategies used:

- [ ] Destratification
- [ ] Aeration
- [x] Hypolimnetic Withdrawal
- [ ] Dilution
- [ ] Other Chemical Trtmt
- [ ] Dredging
- [ ] Wetlands
- [ ] Biomanipulation
- [ ] Copper Sulfate/Cuprine Treatment
- [ ] Alum Treatment
- [x] Filtration
- [ ] Other (specify):

Properties of Open Storage Reservoir(s):

Reservoir Name	Capacity (acre-feet)	Surface Area (acres)	Annual Usage (acre-feet)	Maximum Depth (feet)	Mean Depth (feet)

Water quality monitoring program conducted at:

- [x] Treatment Plant Effluent
- [x] Distribution System
- [x] Storage Reservoirs

Types of water quality parameters monitored for each location used above:

Parameters	Treatment Plant Effluent	Distribution System	Storage Reservoirs
Turbidity	X	X	X
Suspended Solids	X	X	X
EC			
pH	X	X	X
Dissolved Oxygen			
Temperature	X	X	X
Algae Counts			
Algae Identification			
Chlorine Residual	X	X	X
Color			
Total Coliform			
Phosphorus	X	X	X
TKN	X	X	X
Nitrate + Nitrite	X	X	X
Silicate			
Others (list)	Fecal Coliform	Fecal Coliform	Fecal Coliform
	Chloride	Chloride	Chloride
	TN	TN	TN

Frequency of water quality sampling in the storage reservoir(s):

- [ ] Weekly
- [ ] Monthly
- [x] Quarterly
- [ ] In Response to Problems
- [ ] Other (specify):

Water quality problems experienced:

- [x] Algae blooms
- [ ] Odor
- [ ] Turbidity
- [ ] Other (specify):

IMPACT OF SURFACE STORAGE ON RECLAIMED WATER: SEASONAL & LONG-TERM
WERF RFP NO. 99-PUM-4

Survey Reponse for:	***South Bay Water Recycling***
Information Received:	***04/06/00***

Water quality management strategies used:

- [ ] Destratification
- [ ] Aeration
- [ ] Hypolimnetic Withdrawal
- [ ] Dilution
- [ ] Other Chemical Trtmt
- [ ] Dredging
- [ ] Wetlands
- [ ] Biomanipulation
- [ ] Copper Sulfate/Cuprine Treatment
- [ ] Alum Treatment
- [ ] Filtration
- [ ] Other (specify):

Properties of Open Storage Reservoir(s):

Reservoir Name	Capacity (acre-feet)	Surface Area (acres)	Annual Usage (acre-feet)	Maximum Depth (feet)	Mean Depth (feet)

Water quality monitoring program conducted at:

- [x] Treatment Plant Effluent
- [x] Distribution System
- [ ] Storage Reservoirs

Types of water quality parameters monitored for each location used above:

Parameters	Treatment Plant Effluent	Distribution System	Storage Reservoirs
Turbidity	X		
Suspended Solids	X		
EC	X		
pH	X		
Dissolved Oxygen	X		
Temperature	X		
Algae Counts			
Algae Identification			
Chlorine Residual	X		
Color			
Total Coliform	X		
Phosphorus			
TKN			
Nitrate + Nitrite			
Silicate			
Others (list)	see attachment		
	from survey		

Frequency of water quality sampling in the storage reservoir(s):

- [ ] Weekly
- [x] Monthly
- [ ] Quarterly
- [ ] In Response to Problems
- [ ] Other (specify):

Water quality problems experienced:

- [ ] Algae blooms
- [ ] Odor
- [x] Turbidity
- [ ] Other (specify):

IMPACT OF SURFACE STORAGE ON RECLAIMED WATER: SEASONAL & LONG-TERM
WERF RFP NO. 99-PUM-4

Survey Reponse for: ***Tucson Water Treatment Division***
Information Received: ***04/15/00***

Water quality management strategies used:

- [ ] Destratification
- [x] Aeration
- [ ] Hypolimnetic Withdrawal
- [ ] Dilution
- [x] Other Chemical Trtmt
- [ ] Dredging
- [ ] Wetlands
- [ ] Biomanipulation
- [ ] Copper Sulfate/Cuprine Treatment
- [ ] Alum Treatment
- [ ] Filtration
- [ ] Other (specify):

Properties of Open Storage Reservoir(s):

Reservoir Name	Capacity (acre-feet)	Surface Area (acres)	Annual Usage (acre-feet)	Maximum Depth (feet)	Mean Depth (feet)

Water quality monitoring program conducted at:

- [x] Treatment Plant Effluent
- [ ] Distribution System
- [x] Storage Reservoirs

Types of water quality parameters monitored for each location used above:

Parameters	Treatment Plant Effluent	Distribution System	Storage Reservoirs
Turbidity	X		
Suspended Solids	X		
EC			
pH	X		
Dissolved Oxygen	X		X
Temperature	X		X
Algae Counts			
Algae Identification			
Chlorine Residual	X		X
Color			
Total Coliform	X		X
Phosphorus	X		
TKN	X		
Nitrate + Nitrite	X		
Silicate	X		
Others (list)	Fecal Coliform		Fecal Coliform
	Ascaris		
	Cryptosporidium		
	Giardia		

Frequency of water quality sampling in the storage reservoir(s):

- [x] Weekly
- [ ] Monthly
- [ ] Quarterly
- [x] In Response to Problems
- [ ] Other (specify):

Water quality problems experienced:

- [ ] Algae blooms
- [x] Odor
- [x] Turbidity
- [x] Other (specify): Loss of Cl2 residual

IMPACT OF SURFACE STORAGE ON RECLAIMED WATER: SEASONAL & LONG-TERM
WERF RFP NO. 99-PUM-4

Survey Reponse for: *Vallecitos Water District*
Information Received: *04/11/00*

Water quality management strategies used:

- [ ] Destratification
- [x] Aeration
- [ ] Hypolimnetic Withdrawal
- [ ] Dilution
- [ ] Other Chemical Trtmt
- [ ] Dredging
- [ ] Wetlands
- [ ] Biomanipulation
- [x] Copper Sulfate/Cuprine Treatment
- [ ] Alum Treatment
- [x] Filtration
- [ ] Other (specify):

Properties of Open Storage Reservoir(s):

Reservoir Name	Capacity (acre-feet)	Surface Area (acres)	Annual Usage (acre-feet)	Maximum Depth (feet)	Mean Depth (feet)
Mahr Reservoir	166				

Water quality monitoring program conducted at:

- [x] Treatment Plant Effluent
- [ ] Distribution System
- [x] Storage Reservoirs

Types of water quality parameters monitored for each location used above:

Parameters	Treatment Plant Effluent	Distribution System	Storage Reservoirs
Turbidity	X		
Suspended Solids	X		
EC	X		
pH	X		
Dissolved Oxygen			
Temperature			
Algae Counts			
Algae Identification			
Chlorine Residual	X		
Color			
Total Coliform	X		
Phosphorus			
TKN			
Nitrate + Nitrite			
Silicate			
Others (list)	Fluoride	TDS	
	Chloride	% sodium	
	Sulfate	BOD	
	Boron		
	Iron		
	Manganese		

Frequency of water quality sampling in the storage reservoir(s):

- [ ] Weekly
- [ ] Monthly
- [ ] Quarterly
- [ ] In Response to Problems
- [x] Other (specify): When Mahr reservoir is in use

Water quality problems experienced:

- [x] Algae blooms
- [x] Odor
- [ ] Turbidity
- [ ] Other (specify):

IMPACT OF SURFACE STORAGE ON RECLAIMED WATER: SEASONAL & LONG-TERM
WERF RFP NO. 99-PUM-4

Survey Reponse for: *City of Vernon, B.C.*
Information Received: *03/23/00*

Water quality management strategies used:

- [ ] Destratification
- [X] Aeration
- [X] Hypolimnetic Withdrawal
- [ ] Dilution
- [ ] Other Chemical Trtmt
- [ ] Dredging
- [ ] Wetlands
- [ ] Biomanipulation
- [ ] Copper Sulfate/Cuprine Treatment
- [ ] Alum Treatment
- [ ] Filtration
- [ ] Other (specify):

Properties of Open Storage Reservoir(s):

Reservoir Name	Capacity (acre-feet)	Surface Area (acres)	Annual Usage (acre-feet)	Maximum Depth (feet)	Mean Depth (feet)
David Mackay	8000	190	3750	75	50

Water quality monitoring program conducted at:

- [X] Treatment Plant Effluent
- [X] Distribution System
- [X] Storage Reservoirs

Types of water quality parameters monitored for each location used above:

Parameters	Treatment Plant Effluent	Distribution System	Storage Reservoirs
Turbidity		X	
Suspended Solids	X	X	
EC	X	X	
pH	X	X	
Dissolved Oxygen			
Temperature			
Algae Counts			
Algae Identification			
Chlorine Residual	X	X	
Color			
Total Coliform	X	X	
Phosphorus		X	
TKN		X	
Nitrate + Nitrite		X	
Silicate			
Others (list)			

Frequency of water quality sampling in the storage reservoir(s):

- [ ] Weekly
- [ ] Monthly
- [ ] Quarterly
- [ ] In Response to Problems
- [ ] Other (specify):

Water quality problems experienced:

- [ ] Algae blooms
- [ ] Odor
- [ ] Turbidity
- [ ] Other (specify):

IMPACT OF SURFACE STORAGE ON RECLAIMED WATER: SEASONAL & LONG-TERM
WERF RFP NO. 99-PUM-4

Survey Reponse for: *Oso Reservoir*
Information Received: *09/25/01*

Water quality management strategies used:

- [ ] Destratification
- [x] Aeration
- [ ] Hypolimnetic Withdrawal
- [ ] Dilution
- [ ] Other Chemical Trtmt
- [ ] Dredging
- [ ] Wetlands
- [ ] Biomanipulation
- [ ] Copper Sulfate/Cuprine Treatment
- [ ] Alum Treatment
- [ ] Filtration
- [ ] Other (specify):

Properties of Open Storage Reservoir(s):

Reservoir Name	Capacity (acre-feet)	Surface Area (acres)	Annual Usage (acre-feet)	Maximum Depth (feet)	Mean Depth (feet)
not available					

Water quality monitoring program conducted at:

- [x] Treatment Plant Effluent
- [ ] Distribution System
- [x] Storage Reservoirs

Types of water quality parameters monitored for each location used above:

Parameters	Treatment Plant Effluent	Distribution System	Storage Reservoirs
Turbidity	X		
Suspended Solids	X		
EC	X		
pH	X		
Dissolved Oxygen			X
Temperature	X		X
Algae Counts			
Algae Identification			
Chlorine Residual	X		
Color			
Total Coliform	X		
Phosphorus			X
TKN			
Nitrate + Nitrite			
Silicate			
Others (list)	provided on separate sheets		

Frequency of water quality sampling in the storage reservoir(s):

- [ ] Weekly
- [ ] Monthly
- [x] Quarterly
- [x] In Response to Problems
- [ ] Other (specify):

Water quality problems experienced:

- [ ] Algae blooms
- [x] Odor
- [ ] Turbidity
- [x] Other (specify): Daphnia "blooms"

Summary of Survey Responses

Number of agencies reporting open storage reservoirs: 11

Total number of open storage reservoirs: 18

Water quality management strategies used:

Destratification: 1
Aeration: 6
Hypolimnetic Withdrawal: 3
Dilution: 2
Other Chemical Trtmt: 2
Dredging: 0
Wetlands: 0
Biomanipulation: 0
Copper Sulfate/Cuprine Treatment: 1
Alum Treatment: 0
Filtration: 3
Other: 2

Range of properties for open storage reservoirs:

Capacity: 2 - 8000 acre-feet (0.65 - 2,606 MG)
Surface Area: 0.25 - 190 acres
Annual Usage: 365 - 3750 acre-feet (119 - 1,221 MG)
Maximum Depth: 14 - 75 feet
Mean Depth: 14 - 50 feet

Number of water quality programs conducted at:

Treatment Plant Effluent: 11
Distribution System: 5
Storage Reservoirs: 8

Frequency of water quality sampling in the storage reservoirs:

Weekly: 3
Monthly: 3
Quarterly: 2
In Response to Problems: 5
Other: 2

Number of water quality problems experienced:

Algae Blooms: 7
Odor: 5
Turbidity: 4
Other: 4

Types of water quality parameters monitored for each location used above:

Parameters	Treatment Plant Effluent	Distribution System	Storage Reservoirs
Turbidity	9	3	2
Suspended Solids	10	3	2
EC	7	2	1
pH	10	3	2
Dissolved Oxygen	3	0	4
Temperature	5	1	4
Algae Counts	0	0	2
Algae Identification	0	0	1
Chlorine Residual	10	4	4
Color	1	1	1
Total Coliform	9	2	3
Phosphorus	5	2	2
TKN	6	2	2
Nitrate + Nitrite	5	2	1
Silicate	1	0	0
Arsenic	1	0	0
Barium	1	0	0
Cadmium	1	0	0
Chromium	1	0	0
Lead	1	0	0
Mercury	1	0	0
Selenium	1	0	0
Silver	1	0	0
Cyanide	1	0	0
Fluoride	2	0	0
Radioactivity	1	0	0
Sulfate	1	0	0
Boron	1	0	0
TDS	1	0	0
% Sodium	1	0	0
BOD	1	0	0
Iron	1	0	0
Manganese	1	0	0
Ascaris	1	0	0
Cryptosporidium	1	0	0
Giardia	1	0	0
Fecal coliform	2	1	2
Chloride	2	1	1
TN	1	1	1

WASTEWATER UTILITY

Alabama
Montgomery Water Works & Sanitary Sewer Board

Alaska
Anchorage Water & Wastewater Utility

Arizona
Gila Resources
Glendale, City of, Utilities Department
Mesa, City of
Phoenix Water Services Department
Pima County Wastewater Management

Arkansas
Little Rock Wastewater Utility

California
Calaveras County Water District
Central Contra Costa Sanitary District
Contra Costa Water District
Crestline Sanitation District
Delta Diablo Sanitation District
Dublin San Ramon Services District
East Bay Municipal Utility District
El Dorado Irrigation District
Escondido, City of
Fairfield-Suisun Sewer District
Irvine Ranch Water District
Las Virgenes Municipal Water District
Lodi, City of
Los Angeles, City of
Los Angeles County, Sanitation Districts of
Napa Sanitation District
Orange County Sanitation District
Palo Alto, City of
Riverside, City of
Sacramento Regional County Sanitation District
San Diego Metropolitan Wastewater Department, City of
San Francisco, City & County of
San Jose, City of
Santa Barbara, City of
Santa Rosa, City of
South Bayside System Authority
South Orange County Wastewater Authority
Union Sanitary District
West Valley Sanitation District

Colorado
Boulder, City of
Colorado Springs, City of
Littleton/Englewood Water Pollution Control Plant
Metro Wastewater Reclamation District, Denver

Connecticut
New Haven, City of, WPCA

District of Columbia
District of Columbia Water & Sewer Authority

Florida
Broward, County of
Fort Lauderdale, City of
Gainesville Regional Utilities
JEA
Kissimmee, City of, Department of Water Resources
Miami-Dade Water & Sewer Authority
Orange County Utilities Department
Orlando, City of
Reedy Creek Improvement District
Seminole County Environmental Services
St. Petersburg, City of
Stuart Public Utilities
Tallahassee, City of
Tampa, City of
West Palm Beach, City of

Georgia
Atlanta Department of Watershed Management
Augusta, City of
Clayton County Water Authority
Cobb County Water System
Columbus Water Works
Fulton County
Gwinnett County Department of Public Utilities
Macon Water Authority
Savannah, City of

Hawaii
Honolulu, City and County of

Idaho
Pocatello, City of

Illinois
American Bottoms Wastewater Treatment Plant
Dupage, County of
Greater Peoria Sanitary District
Kankakee River Metropolitan Agency
Metropolitan Water Reclamation District of Greater Chicago
North Shore Sanitary District
Wheaton Sanitary District

Indiana
Fort Wayne, City of

Iowa
Ames, City of
Des Moines Metro Wastewater Reclamation Authority

Kansas
Johnson County Unified Wastewater Districts
Unified Government of Wyandotte County/Kansas City, City of

Kentucky
Louisville & Jefferson County Metropolitan Sewer District

Louisiana
Sewerage & Water Board of New Orleans

Maine
Bangor, City of
Portland Water District

Maryland
Anne Arundel County Bureau of Utility Operations
Howard County Department of Public Works
Washington Suburban Sanitary Commission

Massachusetts
Boston Water & Sewer Commission
Upper Blackstone Water Pollution Abatement District

Michigan
Ann Arbor, City of
Detroit, City of
Grand Rapids, City of
Holland Board of Public Works
Lansing, City of
Owosso Mid-Shiawassee County WWTP
Saginaw, City of
Wayne County Department of Environment
Wyoming, City of

Minnesota
Rochester, City of

Missouri
Independence, City of
Kansas City Missouri Water Services Department
Little Blue Valley Sewer District
Metropolitan St. Louis Sewer District

Nebraska
Lincoln Wastewater System

Nevada
Henderson, City of

New Jersey
Passaic Valley Sewerage Commissioners

New York
New York City Department of Environmental Protection

North Carolina
Cary, Town of
Charlotte/Mecklenburg Utilities
Durham, City of
Metropolitan Sewerage District of Buncombe County
Orange Water & Sewer Authority

Ohio
Akron, City of
Butler County Department of Environmental Services
Columbus, City of
Metropolitan Sewer District of Greater Cincinnati
Northeast Ohio Regional Sewer District
Summit, County of

Oklahoma
Tulsa, City of

Oregon
Clean Water Services
Eugene/Springfield Water Pollution Control
Water Environment Services

Pennsylvania
Philadelphia, City of
University Area Joint Authority, State College

South Carolina
Charleston Commissioners of Public Works
Mount Pleasant Waterworks & Sewer Commission
Spartanburg Sanitary Sewer District

Tennessee
Cleveland, City of
Knoxville Utilities Board
Murfreesboro Water & Sewer Department
Nashville Metro Water Services

Texas
Amarillo, City of
Austin, City of
Dallas Water Utilities
Denton, City of
El Paso Water Utilities
Fort Worth, City of
Gulf Coast Waste Disposal Authority
Houston, City of
San Antonio Water System
Trinity River Authority

Utah
Salt Lake City Corporation

Virginia
Alexandria Sanitation Authority
Arlington, County of
Fairfax County Virginia
Hampton Roads Sanitation District
Henrico, County of
Hopewell Regional Wastewater Treatment Facility
Loudoun County Sanitation Authority
Lynchburg Regional WWTP
Prince William County Service Authority
Richmond, City of

Washington
Edmonds, City of
Everett, City of
King County Department of Natural Resources
Seattle Public Utilities
Yakima, City of

Wisconsin
Green Bay Metro Sewerage District
Madison Metropolitan Sewerage District
Milwaukee Metropolitan Sewerage District
Racine, City of
Sheboygan Regional Wastewater Treatment
Wausau Water Works

Australia
South Australian Water Corp.
Sydney Water Corp.
Water Corp. of Western Australia

Canada
Greater Vancouver Regional District
Toronto, City of, Ontario

Mexico
Servicios de Agua y Drenaje de Monterrey, I.P.D.

United Kingdom
Yorkshire Water Services Limited

STORMWATER UTLITY

California
Los Angeles, City of, Department of Public Works
Monterey, City of
Santa Rosa, City of

Colorado
Boulder, City of

Georgia
Griffin, City of

Kansas
Overland Park, City of

Kentucky
Louisville & Jefferson County Metropolitan Sewer District

Maine
Portland Water District

North Carolina
Cary, Town of
Charlotte, City of, Stormwater Services

Pennsylvania
Philadelphia, City of

STATE

Fresno Metropolitan Flood Control District, Calif.
Urban Drainage & Flood Control District, Colo.

CORPORATE

ADS Environmental Services
The ADVENT Group Inc.
Alan Plummer & Associates
Alden Research Laboratory
Alpine Technology Inc.
Aquateam–Norwegian Water Technology Centre A/S
BaySaver Inc.
BioVir Inc.
Black & Veatch
Boyle Engineering Corporation
BPR CSO
Brown & Caldwell
Burns & McDonnell
CABE Associates Inc.
The Cadmus Group
Camp Dresser & McKee Inc.
Carollo Engineers Inc.
Carpenter Environmental Associates Inc.
CDS Technologies Inc.
Chemtrac Systems Inc.
CH2M HILL
Clancy Environmental Consultants Inc.
Damon S. Williams Associates, LLC
David L. Sheridan, P.C.
Earth Tech Inc.
Ecolab Water Care Services
EMA Inc.
The Eshelman Company Inc.
Finkbeiner, Pettis, & Strout Inc.
Frontier Geosciences Inc.
ftn Associates Inc.
Gannett Fleming Inc.
GE Betz
Golder Associates Inc.
Greeley and Hansen LLC
The HACH Company
Hazen & Sawyer, P.C.
HDR Engineering Inc.
HNTB Corporation
HydroQual Inc.
Insituform Technologies Inc.
Institute for Environmental Technology & Industry, Korea
Jacobson Helgoth Consultants Inc.
Jason Consultants Inc.
Jordan, Jones, & Goulding Inc.
KCI Technologies Inc.
Kelly & Weaver, P.C.
Kennedy/Jenks Consultants
Komline Sanderson Engineering Corporation
Lawler, Matusky & Skelly Engineers, LLP
Limno-Tech Inc.
Lombardo Associates Inc.
Malcolm Pirnie Inc.
McKim & Creed
MEC Analytical Systems Inc.
MWH
New England Organics
Odor & Corrosion Technology Consultants Inc. (OCTC)
ONDEO Degremont Inc.
Oswald Green, LLC
PA Government Services Inc.
Parametrix Inc.
Parsons
Post, Buckley, Schuh & Jernigan
R&D Engineering/Conestoga Rover & Associates
The RETEC Group
R.M. Towill Corporation
Ross & Associates Ltd.
Royce Technologies
SAIC Maritime Technical Group
Sear-Brown
Sierra Environmental Services
SYNAGRO
Tetra Tech Inc.
Trojan Technologies Inc.
URS Corporation
USFilter NATC
Vortechnics Inc.
Wade-Trim Inc.
Weston Solutions Inc.
Woodard & Curran
WRc/D&B, LLC
WWETCO, LLC

INDUSTRY

American Electric Power
ChevronTexaco Energy Research & Technology Company
The Coca-Cola Company
Dow Chemical Company
DuPont Company
Eastman Kodak Company
Merck & Company Inc.
ONDEO Services
Procter & Gamble Company
PSEG Services Corp.
Severn Trent Services Inc.
Shell Global Solutions
Thames Water Plc
United Water Services LLC

Note: List as of 5/12/03

WERF Product Order Form

As a benefit of joining the Water Environment Research Foundation, subscribers are entitled to receive one complimentary copy of all final reports and other products. Additional copies are available at cost (usually $10). To order your complimentary copy of a report, please write "free" in the unit price column. WERF keeps track of all orders. If the charge differs from what is shown here, we will call to confirm the total before processing.

Name ______________________ Title ______________________

Organization ______________________

Address ______________________

City ______________ State ________ Zip Code ________ Country ________

Phone ______________ Fax ______________ Email ______________

Stock #	Product	Quantity	Unit Price	Total
			Postage & Handling	
			VA Residents Add 4.5% Sales Tax	
			Canadian Residents Add 7% GST	
			TOTAL	

Method of Payment:

(All orders must be prepaid.)

❑ Check or Money Order Enclosed

❑ Visa ❑ Mastercard ❑ American Express

Account No. ______________ Exp. Date ________

Signature ______________________

Shipping & Handling:

Amount of Order	United States	Canada & Mexico	All Others
Up to but not more than:	Add:	Add:	Add:
$20.00	$5.00*	$8.00	50% of amount
30.00	5.50	8.00	40% of amount
40.00	6.00	8.00	
50.00	6.50	14.00	
60.00	7.00	14.00	
80.00	8.00	14.00	
100.00	10.00	21.00	
150.00	12.50	28.00	
200.00	15.00	35.00	
More than $200.00	Add 20% of order	Add 20% of order	

*minimum amount for all orders

Note: Please make checks payable to the Water Environment Research Foundation.

To Order (Subscribers Only):

Log on to www.werf.org and click on the "Product Catalog."

Phone: (703) 684-2470
Fax: (703) 299-0742.

WERF
Attn: Subscriber Services
601 Wythe Street
Alexandria, VA 22314-1994

To Order (Non-Subscribers):

Non-subscribers may be able to order WERF publications either through WEF (www.wef.org) or IWAP (www.iwapublishing.com). Visit WERF's website at www.werf.org for details.